Meet the editors

Prof. Nesibe Ebru Kafkas graduated from the Faculty of Agriculture, Department of Horticulture, University of Çukurova, Türkiye, where she has been a researcher and lecturer since 1996. She has experience in fruit breeding and fruit metabolomics, chromatography, and in vitro techniques. She is the author of more than 100 scientific papers and an editor for various scientific journals. She has been the Turkish coordinator for national and international EU projects such as HORIZON 2020, Cost, and EU Prima. She is also a member of the International Society of Horticultural Science.

Prof. Hüseyin Çelik is an academician in the Faculty of Agriculture, Department of Horticulture, Ondokuz Mayıs University, Türkiye. His research interests include the breeding, propagation, and performance of berry fruits, as well as berry characteristics, antioxidants, and chemical compositions. He is currently working on blueberry and foxy grape (Vitis labrusca) breeding. Prof. Çelik is participating in several international projects with Germany and Slovenia. His work is funded in part by the Scientific and Technological Research Institution of Turkey (TUBITAK) and national private investors.

Contents

Preface

This book discusses the biochemical contents, human health benefits, disease control methods, and production of berry fruits. It also examines quality losses in post-harvest processes as well as how biotic and abiotic stresses and diseases caused by climate change affect these fruits.

Berry fruits are a good source of antioxidants, containing vitamins and flavonoids such as anthocyanins and flavonols. They can prevent the increase of free radicals that cause oxidative stress and diseases such cancer and are preferably consumed as dietary fruit, especially by health-conscious consumers. For these reasons, there has been an increasing trend toward the consumption of berry fruits in recent years. In terms of processing technology, berry fruits are a raw material source for many food sectors. Written by experts, this book addresses these and other aspects of berry fruits. The book is an essential resource for virtually everyone involved in berry fruit research and development, including academics, industry professionals, farmers who produce berry fruits commercially, and consumers.

I would like to thank Assistant Editor İlbilge Oğuz and Author Service Manager Zrinka Tomicic at IntechOpen for their courteous help and dedicated efforts, without which this book would never have been published.

Nesibe Ebru Yaşa Kafkas
Professor,
Faculty of Agriculture,
Department of Horticulture,
University of Çukurova,
Adana, Türkiye

Hüseyin Çelik
Professor,
Faculty of Agriculture,
Department of Horticulture,
University of Ondokuz Mayıs,
Samsun, Türkiye

Section 1

Ecology Requests of Berries

Chapter 1

Preferable Berry Fruits for Tolerance to Global Climate Change and Dry Conditions

İlbilge Oğuz, Halil İbrahim Oğuz, Şule Hilal Attar, Duygu Ayvaz Sönmez, Hüseyin Çelik and Nesibe Ebru Yaşa Kafkas

Abstract

Global climate change and possible drought scenarios have forced researchers, breeders and producers to create new plant patterns that will adapt to changing climate and soil conditions for production of horticultural plants in the future. Here, the most important topic is the shortening of the physiological growth period of plants due to abiotic stress. In other words, reductions experienced in both cooling needs and maturation periods cause negative impacts on flowering times and amounts, and this causes significant loss of yield. In recent years, the production of berry fruits that will adapt to drought conditions has attracted the attention of breeders and producers. The aim of this study is to discuss in detail the possibilities of producing berry fruits that are resistant to drought and negative climate conditions and to present research results and recommendations about this topic. In this study, the production opportunities in arid and negative climate conditions for the berry fruits of strawberry (*Fragaria vesca* L.), mulberry (*Morus* spp.), fig (*Ficus carica* L.), blackberry (*Rubus fruticosus* L.), chokeberry (*Aronia melanocarpa* L.), rosehip (*Rosa canina* L.), raspberry (*Rubus idaeus* L.) and blueberry (*Vaccinium corymbosum* L.) were researched and recommendations are made about production methods for some varieties and types resistant to drought in berry fruit cultivation.

Keywords: climate change, aridity, berry fruit, adaptation, production models

1. Introduction

Agricultural activity, an ecosystem linked to nature, is one of the sectors most affected by global climate change. Climate change occurring as a result of greenhouse gas emissions negatively affects all agricultural activities. Additionally, fruit cultivation is the agricultural activity most sensitive and most intensely impacted by global warming and climate change. As a result, fruit trees are negatively affected by unsuitable climate conditions during winter rest, flowering, budding and fruiting periods and large economic losses occur. It is predicted that climate change will cause global food crises, that food crises will become more severe with climate change and that small-scale farmers will be the first and most severely impacted group as a result of climate change. Among agricultural products, the agricultural products most negatively impacted by climate change and greenhouse gas emissions come from horticultural plants. One of the most negative physiological responses to climate change is experienced in the growth and development periods of fruit and vegetable production. These effects will create negative outcomes for the growth and development of horticultural plants with water loss from soil due to temperature increases. As a result, within the framework of ecological change, it is mandatory to perform breeding and adaptation studies of species and varieties suitable for climate change on a product basis [1].

The IPCC predicts that in the period from 2070 to 2100, the mean temperature in Türkiye may increase from 3 to 7°C. This increase will be higher in warm seasons compared to cold seasons, with temperatures in the summer months predicted to be 4.5°C higher while temperatures in winter months will be 3.5°C higher. According to a more pessimistic scenario, the expected increases may reach 7°C in the summer months and 4.5°C in the winter. According to the results of regional climate models, a clear reduction is expected in rainfall in the west and south regions where the Mediterranean climate is dominant, while an increase in rainfall is expected in the Black Sea region. Due to increasing temperature and reducing rainfall, the severity, frequency and duration of drought events are expected to significantly increase [2]. Drought events, frequently experienced due to anthropogenic climate change, the resulting increasing atmospheric temperatures, and changes in rainfall will significantly damage global agricultural systems and agricultural production [3, 4]. Due to the complicated climate structure of Türkiye, it appears to be one of the countries that will be most affected by climate change due to global warming [5, 6]. Additionally, as is known, climate change will impact different regions in different ways and at varying scales as Türkiye is surrounded on three sides by seas, has variable topography and orographic features. For example, just as the Southeast and Central Anatolian regions will be under threat of further desertification due to temperature increases, arid and semiarid regions and semi-humid Aegean and Mediterranean regions without adequate water will be more impacted [5]. It is considered that the significant reductions in rainfall in the Seyhan River basin in 2070–2100 and changes in the amount of snowfall and melt times will change the seeding/planting times for basic products like wheat and corn, and more importantly will change the planting localities [7]. In Türkiye, there are serious concerns related to the increases that will occur in summer and winter temperatures, reductions in rainfall and agricultural production, and increases in sea level [8].

In light of this knowledge, undoubtedly the negative climate conditions will take their toll on berry fruit production. However, berry fruit generally do not cover a large area in terms of production area, while the high possibility for horizontal or vertical agriculture for berry fruits like strawberry, raspberry, gooseberry and blueberry, especially, has attracted the attention of producers. The product with highest cultivation in berry fruit production around the world is the strawberry. For global blueberry production, the USA is in first place producing 308,760 tons, followed by Canada in second place with 176,127 tons, Peru in third place with 142,427 tons and Spain in fourth place with 53,380 tons. For global raspberry production, Russia is in first place producing 174,000 tons, followed by Mexico second with 128,848 tons, Poland third with 150,000 tons, Serbia fourth with 120,058 tons and USA in fifth place with 102,510 tons production. Türkiye has 8% share of global berry fruit production and it was reported that 24.92% of this is strawberries, while 25.18% is other berry fruit species [9]. As a result, the convenience of producing berry fruit will increase their preferability in the struggle with global warming and drought. Additionally, berry fruits generally have bush and spreading forms and just as they can be cultivated in very large areas, they can also be cultivated in narrow and small areas, which makes these plants more advantageous. Berry fruits grow on small fruit plants with different colors like blue, purple and red. Berry fruits with more common production around the world in recent years may be listed as chokeberry, barberry blueberry, cranberry, blackberry, raspberry, white, red or black currant, and strawberry. Additionally, the shelf-life of berries is short and their use in processed food like fresh fruit, fruit juice, jam and drinks is limited. Additionally, the majority of berry fruits require different storage methods in processes like drying and packaging to preserve the quality of fruit during storage [10]. Berry fruits contain flavonoids (anthocyanins, flavonols and flavanols like concentrated tannins or proanthocyanidins), hydrolyzable tannins (ellagitannins and gallotannins), phenolic acids (hydroxybenzoic and hydrocinnamic acids, chlorogenic acid), and high levels of polyphenols like stilbenoids and lignans [11, 12].

Berry fruits grew wild in forests in the first years and were cultivated mainly as fence plants along field boundaries. It is known that cultured varieties of berry fruits were derived from soft- and hard-seeded fruits toward the end of the 16th century. Currently berry fruits are very commonly grown, especially in Europe; however, apart from strawberries, cultivation of the others is still new in Türkiye. As berry fruits are very rich in aroma agents, they are used at broad scale mainly in the marmalade, jam, fruit juice and drink industries [9, 13, 14]. Additionally, berry fruits contain phytochemicals like phenolic acids (hydroxybenzoic and hydroxycinnamic acid), flavonols (quercetin, kaempferol, myricetin), flavanols (catechins and epigallocatechin [EC]), anthocyanins (cyanidin glycosides and pelargonidin glucosides) and flavonoids like tangilanins. Anthocyanins responsible for the color of the fruit are found in abundant amounts in the skin of the fruit [15–20]. Due to these rich bioactive components, there is increasing demand for berry fruits in recent years. Hence, berry fruits are products with high added value and economic value. While the wild berry fruits found in the natural flora of Türkiye are collected from forest areas and sold for fresh consumption in local markets, cultivated species are processed or frozen for use in products like fruit juice, jam, jelly, puree and ice cream due to being seasonal [21–24].

Additionally, due to the wealth of active components in berry fruits, several studies have been performed as consumption of berry fruits like mulberry, chokeberry, barberry blueberry, cranberry, blackberry, raspberry, white, red or black currents and strawberry is very beneficial in terms of health [25–28]. Additionally, in recent years, drying processes have been performed for berry fruits, as for many types of fruit. New technologies used for drying processes contribute to the struggle against global warming and climate change. For export of fresh fruit, high cost storage and transport processes are implemented to prevent degradation of fruit during storage and transport. This process increases both costs and energy consumption as input costs increase and more energy is used. At this point, trade in dried fruit has gained more importance for market presentation in recent years. For example, for quality features of dried fruit, drying conditions and techniques used require great care due to fruit being very sensitive to degradation of phytochemical content like anthocyanins and phenolic compounds [10, 12]. Classic drying processes generally cause losses of the functional properties and antioxidant activities of fruit. As a result, the drying process conditions and methods used are known to significantly affect the bioactive compounds in the fruit, linked to the fruit species. Additionally, fruit dried with the lyophilized drying method, for example, generally do not have any changes in color, nutrients, vitamin values, aroma, odor and fruit shape. This and similar methods are considered to contribute to methods in the struggle against global climate change and drought in the future, though costs are high at the moment [12, 29]. Additionally, it is known that climate factors have very significant effects on biochemical properties of fruit. For example, bioclimatic data about atmospheric forcing on the vine in the long term identified different climate variables in certain places in the long-term period had significant effects on the growth cycle of vines, biomass production and grape properties based on temperature and radiation intensities and durations in a spatial and temporal sense in viticulture [30, 31]. Some of the most important factors affecting fruit production and quality of berry fruits are extreme climate conditions during winter rest, flowering, budding and fruiting periods [32]. In fact, it was reported that budding would be delayed as a result of not meeting cooling needs linked to temperature increases for viticulture in the Margaret River region in Australia [33]. To better prepare for possible impacts of climate change, it was reported that it was necessary to perform more research about breeding varieties with lower cooling requirements, developing rootstocks that can cope with inadequate winter cold and understanding the responses of products to temperature better [34]. For example, in Downeast, Maine region of the USA, research about wild blueberry fruits reported that summer temperatures significantly increased over 40 years in natural blueberry fields due to climate change and this situation caused farmers to experience significant economic losses [35].

Every year in different regions and provinces significant losses are experienced in many agricultural sectors due to the negative impact of climate change on drought, hail, heavy rainfall, floods, frost, storms and other abiotic stresses. Additionally, changing climate events, unbalanced rainfall and high-low temperature regimes threaten all sectors of agricultural production [36]. Climate change causes a reduction in fruit and vegetable production from annual and perennial horticultural plants for agricultural production, shortening the growth duration of plants due to negative effects on growth and development of plants linked to terminal heat stress, especially. The results will provide a beneficial guide for planning vine development in the future in Serbia. However, contrary to what is known, this research reported that cold damage, usual to date, may be reduced in cold regions, especially. However, it was reported that there is a need for more detailed studies and to develop concrete adaptable strategies to clearly say that there will be positive effects for viticulture in Serbia. Additionally, it was reported that research in the future should use multi-model and multi-scenario climate modeling with the holistic approach, identify the accuracy of data obtained from this modeling and measure the correlation between grape quality and yield by revealing the impacts on vine phenology due to climate change in the region. Other factors that need to be explained include identification of the impacts on plants of other meteorological factors like humidity and wind, in addition to temperature and rainfall; direct effect of increasing atmospheric carbon dioxide concentrations on vines; impacts of soil, topography and other environmental factors on vine products; impacts of climate change on disease and pests; and determining the possible physiological reactions of some vine varieties to predicted climate changes [37]. In many European countries, though strawberries are cultivated in strawberry gardens using plastic cover material as they are early producers and to protect from cold, strawberry cultivation is still performed in open areas. As is known, plastic cover materials are commonly used with the aim of preventing damage from frost in the flowering period in spring and to ensure early yield in strawberry cultivation. Additionally, in many countries in Europe, around the world and in Türkiye, automation in production of berry fruit like strawberry, blueberry, blackberry and mulberry uses steel, wood and similar constructions to create gothic, arc roof, tunnel and glass greenhouses produced from plastic and glass (with shades) appropriate to the geographical and climate conditions suitable for the berry fruit type and variety [38]. The aim of this study is to provide information about which production methods and models should be applied in climate regions for the cultivation of berry fruits linked to temperature, rainfall and aridity, and what the appropriate rehabilitation work is for a variety of rehabilitation, drought and limited irrigation methods considering aridity and vegetation indexes. Information is also given about care and training systems for berry fruits, irrigation, methods against disease and pests, marketing and processing techniques.

2. Cultivation models for berry fruit compatible with aridity and climate change

In this section, the focus is on botanic and ecological needs and nutritional content of berry fruit with easy cultivation and high adaptation to different climate conditions to be able to use them in arid and global climate change scenarios. In general, the ability of individual genotypes of all plants to create distinct phenotypes when exposed to varied environmental situations is referred to as phenotypic plasticity. Environmental influences on the expression and function of genes that regulate plastic characteristics cause these phenotypic variations [39]. As in all fruits, fruit phenotypic properties such as sugars, acids, phenolics, anthocyanins and aroma compounds in berry fruits are significantly affected by variety and environmental conditions. The combination of these phenotypic parameters depends entirely on the stability of fruit traits. Researchers defined this situation as "the tendency to keep the characteristic of a physiological or morphological system constant [40]. Despite the fact that phenotypic plasticity is a significant ecological phenomenon, genetic and molecular mechanisms are insufficient to fully explain it [41]. Moreover, phenotypic variation between species and organisms within the same species can effect variances in gene expression as well as differences in gene structure. However, it is hypothesized that phenotypic plasticity between clones of the same genotype is far more dependent on variable gene expression in different contexts [42]. Recent research has shown the presence of high-throughput expression profiling tools, which making it possible to analyze the genetic mechanism (activity and spatio-temporal features) on a global scale. As a result, the impact of transcriptome plasticity may now be studied directly [42–44]. In order to adapt to climate change, new research and applications continue to be developed in terms of all kinds of plant and animal production techniques. Especially in the production of berry fruits, researches on the choice of soil location, tillage, seedling production from production techniques, planting system, pruning and training systems, changes in the planting environment (soilless culture, horizontal or vertical agriculture, plant nutrition and garden management systems with machine learning content, artificial plant environments) continue at full speed. For this purpose, in a study conducted by [45], they stated that regarding the berry orchard, applications that increase the organic matter level of the soil should definitely be made, the berrys in the garden are different in terms of soil compaction in the management applications should be considered; and for example, different doses should be tried for the soils of different berry species for an economical approach in the barnyard manure applications. Furthermore, they recommended that preliminary soil analyses be performed before establishing berry orchards, and that berry varieties suitable for these soil qualities be favored. For instance, berry types with stronger root structure than other berry types, such as aronia, blackberry, and goji berry, should be favored in semi-arid environments. Besides, except very hot, arid, and foggy environments, blackcurrant growing does not provide significant climatic challenges. It is more cold-resistant than other berry fruits. Black currants are vulnerable to hard winters. Currant cultivation is dangerous in areas prone to spring frost [46]. The discovery of "primocane" types, which take fruit from flower buds grown on one-year shoots, is the most significant breakthrough in blackberry production today. These variations are favoured not because they produce higher-quality items, but because they allow harvest to extend a longer period of time through the use of techniques such as tip picking into breeding systems. In recent years, thornless primocane cultivars have also been developed. Adaptations of these cultivars continue in greenhouse cultivation, particularly along the Mediterranean coast. The private sector is developing healthy and high quality seedling production, variety adaptation and introduction, extension of the harvest period with different breeding techniques, greenhouse, fertigation, correct harvesting and preservation techniques. Planting healthy, suitable seedlings at the correct time is a critical aspect in strawberry production. Strawberry seedlings used in cultivation include frigo, fresh, green, and "plug" seedlings. The frigo seedling is the most often utilized seedling type in strawberry growing in Turkey and many other nations across the world. However, in recent years, the use of frigo seedlings has been completely abandoned, especially for early growing; and the use of fresh and "plug" seedlings has begun. Due to the ecology and planting period during which flower bud production occurs, it is feasible to obtain fruits earlier in culture with these seedlings, and harvest is extended a wider period [47]. Thus, changes are made in the production and breeding techniques mentioned above, taking into account climate change and ecological conditions in the production of berry fruits. Furthermore, numerous production practices, such as mulch treatments, soil processing processes, and irrigation systems, have had to modify as a result of climate change and temperature increases.

The study of species and cultivars suitable for drought and adverse climatic conditions in berry fruits is one of the priority issues of plant breeders all over the world. Honey berry (*Lonicera caerulea* L.) production, for example, is increasing in certain European countries as well as Canada and Japan. The explanation for this is that it is a plant of Russian origin with a remarkable adaptation to severe environmental conditions. Besides its fruit is a sort of berry with a distinct flavor reminiscent of raspberry, blackcurrant, and blueberry. Its fruits range in flavor from bitter to sour to sweet [48]. Honey berry is notable for its early maturity, high ascorbic acid content in its fruits, and bioactive flavonoid content. One of the most important features of this berry is its exceptional frost resistance to cold, its suitability for mechanized harvesting, and its resistance to diseases and pests [49]. Its cultivation is still taking place in Russia, Japan, Poland, the Czech Republic and Canada. There are more kinds bred in Russia, where the population density is high. Tundra, Berry Blue, Indigo Gem, and Borealis were among the major cultivars launched in Canada in the early 2010s. New kinds include the honey bee, Aurora, Boreal Blizzard, Boreal Beauty, and Boreal Beast [50]. Boreal Beauty (*Lonicera caerulea* var. edulis) is a new variety with superior characteristics developed in Canada. Boreal Beauty carries the Russian, Japanese and Kuril gene. It is the most heat tolerant variety. Boreal Beast (*Lonicera caerulea* var. kamtschatica) grows upright in the form of a bush. It is very similar to Boreal Beauty. It is a variety with high tolerance to diseases and pests. It is a highly productive variety tolerant to powdery mildew. As a result, studies are continuing in the breeding of species and varieties that are tolerant to climate change, drought, diseases and pests in berry production in the world.

2.1 Strawberry (*Fragaria x ananassa* L.)

In recent years, strawberry cultivation under protected conditions has increased for the fresh market in European countries. Producers choose very short day varieties in order to reach the market early. The aim of protected cultivation is to protect the strawberry from negative weather conditions. The selection of varieties with low cooling needs and high quality is becoming more common for strawberry cultivation in the winter and spring in the Mediterranean climate, while generally neutral varieties in terms of light needs are chosen for strawberry cultivation in the open air in cool places. Additionally, producers choose summer planting for strawberries rather than spring planting to prevent the effects of cold in continental climates. The use of refrigerated seedlings for summer and autumn strawberry production has significantly increased. This has contributed to the increase in strawberry cultivation in greenhouses and plastic tunnels. As a result, many different production techniques are used to offer fresh strawberries to the market in nearly every season by developing new production models for strawberry production in Europe in parallel with climate change [38]. As is known, the effect of climate change on plants is a topic that requires study by scientists and there is continuous research about this topic.

2.2 Black chokeberry (*Aronia melanocarpa* Michx)

Chokeberry is a plant species with superior adaptability to pedoclimatic conditions (wet habitats, sand dunes, rocky slopes and steep cliffs and extremely overgrown or bare rock outcrops). Additionally, chokeberry plants may be cultivated in marginal areas in terms of nutrients [51]. It is a plant that has the ability to tolerate dry or excessively humid soils [51] and very low temperatures and additionally is resistant to plant diseases and pests [52]. Due to these features, chokeberry is a berry brush that can be very easily used for rehabilitation with plant cover in abandoned areas [51]. The basis of the adaptation success of this plant species is linked to the ability to adjust to the soil pedoclimatic conditions and continuously changing environmental conditions. In this context, a study completed in Romania Research Institute for Fruit Growing Pitesti—Maracineni related to adaptation to specific climate conditions of Melrom and Nero chokeberry varieties used the Biologische Bundesantalt, Bundessortenamt and Chemische Industrie (BBCH) scale for phenological observations. The study calculated the mean flowering and fruiting initiation, flowering and fertilization duration, mean air temperature, total solar radiation and cold and heat accumulation. According to this study, budding began from 28 January – 8 February, bud opening began on 3 March, flowering began from 15 April to 1 May and finished from 27 April to 14 May. Phenological observations obtained a BBCH value of 53–87, while the mean air temperature over 154 days was reported to vary from 6.1 to 36.8°C. At the end of this research, they identified that chokeberry (*Aronia melanocarpa*) was a berry fruit species that was adapted to the temperate continental climate of south Romania [53].

In Türkiye, blueberry studies began in the 2000s in the Black Sea region, and in Rize. A study in Samsun province in Türkiye found the fruit harvest for blueberry varieties began from the end of June to beginning of July and continued until the middle of

August. For extension of the harvest period, they reported the cooling needs of Misty, Ozarkblue, O'Neil, Jubilee and Sharpblue, low height southern blueberry varieties and late maturing rabbit-eye blueberry varieties of Climax, Powderblue, Tifblue and Austin had full flowering in the ecological conditions in Samsun Türkiye. The Misty blueberry variety with earliest low cooling needs fully flowered on 3 March, while for latest full flowering the rabbit-eye blueberry Tifblue variety flowered on 5 April. The earliest leaf buds were found on the Misty variety on 25 January, while the latest leaf buds were present on the rabbit-eye blueberry variety of Tifblue on 28 March. For the fall time, the southern ones were earliest on 7 May for Misty, while for the rabbit-eye variety, the latest occurred on 10 June. The rabbit-eye blueberry varieties had highest fruit yield (1092.3 g/plant, Austin), while the heaviest and largest fruit were obtained from Climax (3.80 g and 20.70 mm), with hardest fruit obtained from the rabbit-eye blueberry variety of Powderblue (3.74 kgF). In conclusion, in the Samsun coastal ecology, the harvest period was earlier for the southern tall blueberry varieties, while the rabbit-eye blueberry varieties were slightly later and could extend the yield period until August [54].

2.3 Raspberry (*Rubus idaeus* L.)

Raspberry is a berry fruit that can be cultivated in different climate conditions by ensuring production conditions, and is suitable for boutique-style production in bush form. The Latin name is *Rubus idaeus*, and the raspberry fruit is in the Rosaceae family. It is a plant with easy cultivation and low cost. It does not require much technical knowledge for rootstock use and production, does not have a grafting problem and is easy to prune. It may provide fruit in a short duration after planting. As the roots of the plant comprise fringe roots with multiple and frequent thin roots, they may reach 1–1.5 m depth in appropriate soil conditions. The roots are perennial but stems last 2 years. As a result, the stems are renewed each year. As it is not very tall, care and harvest are easy. The harvest period may last 4–6 weeks duration. As a result, harvesting is flexible as labor requirements are not focused within a few days. Generally, raspberry cultivation may be successfully performed in areas with abundant sunlight, protected from wind and with adequate soil humidity [55–57]. According to historical records, it has been used as a medicinal plant since 1898. Tea made from the leaves is reported to have good odor and flavor. Additionally, the leaves contain tannin, flavonoside, organic acids and vitamin C, while fruits contain organic acids, sugars, pectin, vitamin C, anthocyanin, volatile and stable oils [58–60]. Dioscorides mentioned that the flowers of the plant mixed with honey were used for eye infections, were good for skin diseases and the fruits facilitated the digestive system in his book called Materia Medica. Raspberry fruits are commonly consumed as fresh fruit, as functional drink and fermented wine due to perfect nutritional properties rich in antioxidant phenolics and with attractive color and tasty flavor [61, 62]. In traditional medicine raspberry is frequently used for treatment of flu-like infections, and is said to be a very popular fruit in Poland and Europe [63]. Raspberry is reported to be commonly cultivated in Türkiye, especially Mount Ida (Kaz Dağları) [64].

2.4 Blackberry (*Rubus fruticosus* L.)

In Türkiye, the importance of blackberry cultivation is increasing due to the suitable ecology for blackberry cultivation, ease of cultivation, giving fruit in a short duration, high yield per unit area, higher price compared to other fruit species and consumption in different forms as food [65, 66]. Blackberry is in the Rosaceae Rubus genus with species assessed within the Rubus subgenus. The basic feature separating blackberry from raspberry is whether the torus (fruit stem) remains on the fruit or not. When collecting blackberry fruit the torus remains on the fruit, while with raspberry the torus remains on the plant and the section where the torus entered the fruit forms an internal cavity. The general Latin name for the fruit is *Rubus fruticosus*. Fruit may weigh 3 to 12 g linked to the variety, while it is actually a cluster of fruitlets consisting of several drupelets, each containing a seed [67, 68]. It may be grown in soils that are inadequate in terms of nutrients. The plant has white or pink flowers from May to August, and has fruits from black to dark purple in color. The blackberry plant may grow to 3 m in height. Fruits are generally consumed fresh, while they are used frozen or freeze dried in the food industry. They are commonly used in cake products, jam, marmalade, fruit juice, ice cream, puree, sweets, concentrate and wine production due to color and aroma. The main production regions for blackberry are countries in North America, Europe, Asia, South America, Oceania, Central America and Africa. Additionally, wild blackberry may negatively affect sales of commercially-grown fruit in some regions [69–72]. The leaves of the plant have astringent features and are used in ethnic medicine for diarrhea, hemorrhoids, wound treatment and to balance blood sugar levels. Additionally, fruit are known to be used as antiseptic for eye infections, and as mouthwash for gum, tonsil and throat infections. Blackberry is a fruit with high antioxidant capacity due to containing high anthocyanins and ellagitannins (ET) with other phenolic agents. Epidemiological and clinical studies show that the consumption of anthocyanins and other flavonoids found in most fruit and vegetables may reduce the risk of obesity, coronary heart disease, degenerative status and a variety of cancer types [15].

2.5 Rosehip (*Rosa canina* L.)

Rosehip is called many different names by the public including wild rose, chillan, crazy rose, rosehip, rose apple, doghip and dogrose. It grows naturally in all regions of Europe and Asia and in Türkiye. In Türkiye, it grows on its own up to 2000 m elevation, on mountain slopes, in the internal openings in forests, on forest edges, in heathlands, and along stream and road edges in areas with abundant sunshine-semi-shade and humic soils [73]. As flowers open in May, June, and July, they are not damaged by late spring frosts. The flowering time is delayed as the elevation increases and the fruit quality increases. Adequate rainfall in the vegetation period increases the size of fruit. In locations with high and abundant sunshine and with south aspect, fruit color and size increase along with the vitamin C content. Rosehip is an important source of income for forest villages in rural areas. Rosehip (Rosa spp.) belongs to the Rosa species in the Rosaoideae subfamily of the Rosaceae family in the Rosales order. Of the 70–100 species of rosehip globally, nearly 25% (27 species) grow in our country [74–77]. Varying according to rosehip species, it is a deciduous bush-like plant that may reach 0.5–4.0 m with vertical and weeping forms, with few or many thorns on stem and branches. The stem and branches have weeping appearance, while most have frequently thorny and creeping, climbing forms. Branches from the stem are thorny and sturdy and only reach 1 cm diameter at 3 years. Leaves comprise 5–11 hairless leaflets, leaflets are 2–4 cm in length, with shiny green color on the tops of the leaves [77]. Rosehip has strong root structure. It has both surface fibrous roots and deep (up to 4 m) taproots. Roots are resistant to disease, pests and difficult conditions. The roots with red color and soft fleshy structure are used in the dye industry. Due to nodules on the roots, they tend to produce root shoots [78].

Flowers are single or collected in umbrella-like bunches, and have light red, pink, yellow, cream or white color. The sepals have round or long egg-like appearance, tips later recline, and then shed or remain on the fruit, depending on the species. The rosehip has hermaphrodite flower structure, with many male and female organs. Flowering occurs from April, May and June linked to species and climate and lasts 15–25 days [75, 77].

2.6 Mulberry (*Morus* spp)

As is known, mulberry is one of the most perfect plants compared to other berry fruits for climate change and global warming conditions in barren areas. This is because care and cultivation conditions are easier compared to other berry fruits, and it is a beneficial berry fruit that does not require much water, is suitable for barren conditions, both fruit and leaves can be used in the food industry, in silkworm production and folk medicine. In spite of Türkiye, with fruit cultivation culture extending through history, being the home of the mulberry and having natural areas of distribution, the genetic potential has not been sufficiently evaluated. This fruit type, with highly superior qualities in terms of fruit quality, is cut only to use the wood leaving many valuable genotypes at risk of being lost [79]. In addition to fresh mulberry consumption, mulberry has significant potential as a processed food product due to nutritional features. Products like mulberry molasses, jam, pestil (fruit leather), mulberry paste, fruit ice cream, mulberry sujuk, vinegar, fruit juice concentrate, and spirits are made in the regions in which it grows. Black mulberry juice has become a very common drink in recent years especially [80]. Mulberry (Morus spp.) is in the Morus genus of the Moraceae family within the Urticales order. The number of species within the Morus genus was reported to be 12 by Freeman [81], 14 by Huo [82], 24 by Koidzumi [83] with 1 subspecies Machii et al. [84], more than 30 by Martin et al. [85] and 68 by Datta [86]. Mulberry is commonly encountered especially in east, west and southeast Asia, southern Europe, south of North America, northwest of South America and some regions in Africa. Mulberry plants grow to 15 m in height. They grow rapidly, with cylindrical, straight and thick trunks, with fractured bark and gray-brown color. The diameter of the crown is 6–8 m, with sparse and ball-like appearance. Roots are fleshy, brittle and fragile. As they age, strong lateral roots develop. As a result, they are resistant to wind. Leaves have stems, arranged in two rows, with rounded base or heart shape, top surface dark and bottom surface lighter green. The leaves generally have a pointed tip, with toothed edges. Shoots are shiny yellow in color and slightly hairy. When shoots are cut, they release a milk-like secretion. Mulberry has male and female flowers on the same tree. Flowers open in April–May. Pollination occurs with the wind direction generally. Mulberry fruit is in the form of a collection of fruitlets (multiple) each forming from a flower on the flower stem [79]. Mature fresh mulberry contain 85–88% water, 7.8–9.2% carbohydrate, 0.4–1.5% protein, 0.4–0.5% fat, 1.1–1.9% free acids, 1.4% fiber and 0.7–0.9% minerals [87]. Additionally, 100 g of fresh mulberry provides 93 calories, containing 0.9 g protein, 19.8 g carbohydrate, 1.1 g fat, 0.9 g fiber, 60 mg calcium, 1.1 mg iron, 0.05 mg vitamin B1, 0.07 mg vitamin B2, 0.2 mg vitamin B3 and 17 mg vitamin C. Additionally, there is 11.0–12.5 mg ascorbic acid, 0.7–0.8 mg nicotinic acid, 7.0–9.0 μg thiamine and 165–179 μg riboflavin within 100 g of fruit [88]. The syrup obtained by boiling black mulberry fruit harvested in the fully mature period with water, sugar or honey mixture is rich in vitamins and minerals, in addition to being recommended as a drink due to blood purifying and antioxidant properties. Additionally, it is stated to be beneficial when used 3–4 times per day as a mouthwash for sore throat and gum infections [89]. Mulberry juice, produced as a health drink in a commercial sense, has become popular in China, Japan and Korea. Original mulberry juice remains fresh for 3 months in cold storage conditions without any added preservatives; the bottled drink preserves its freshness for 12 months at room temperature. Just as fruit can be used fresh for jam and compote and leaves can be used fresh for stuffed leaves, they may be used in a variety of forms by drying. Fruit, leaves and roots can be used to obtain extracts and these are used in a variety of cosmetic products. Just as white dried mulberry, eaten as trail mix, can be used to lower inflammation in blood and to increase breastmilk, it may be assessed for use in late-healing wounds in diabetes patients and for eczema [90].

2.7 Fig (*Ficus carica* L.)

The Ficus genus in the Moracea family includes nearly 600–2000 species. Most of these species are found in the tropics and subtropics and the fruit of only a very small number among these species can be eaten [91]. Among these species, the fig most commonly consumed by people is *Ficus carica* L. The *Ficus carica* L. species has great importance as a source of human nutrition and is the only Ficus species cultivated for fruit [92]. As understood from records dating to the 30th century BC, fig trees were the first trees planted in the world. Though the origin is not fully known, fig trees are thought to have emerged from the productive region of south Arabia in western Asia. At the same time, many wild species are found in the Middle East and Mediterranean regions [92–95]. Another berry tree resistant to climate change and arid conditions is the fig. Fig (*Ficus carica* L.), one of the first cultivated trees in the world, grows in many locations around the world with temperate climate. Figs are eaten both fresh and dried; however, as fresh figs are products that degrade quickly, they are mainly consumed close to the production areas. Figs are nutritional fruit rich in terms of fiber, potassium, calcium and iron. Fresh figs are very vulnerable to physical damage and are sensitive to rot infections after harvest. Conditions before and after harvest improve fruit quality and post-harvest life. At this point, there is great difficulty for plant cultivators, physiologists and post-harvest technology experts to reduce post-harvest losses and develop global fresh fig marketing [96]. In 2022, there was 350 tons of production from 572,472 thousand area with 57.3% of production from Aydın and 19.5% of production from İzmir. The majority of figs produced in Aydın and İzmir are dried due to climate and ecological conditions, with fresh production from other provinces. In 2021, 21.4 thousand tons of fresh fig and 57 thousand tons of dry fig were exported [97]. On a global scale, fig is produced from 299,541 hectares and production was more than 1,348,254.74 tons in 2021. Fig trees are commonly planted in gardens in the Mediterranean region (and similar climates) and have good adjustment to aridity and high temperature. Türkiye and Egypt are the countries with highest fig production representing nearly 50% of global production, followed by Iran, Algeria and Morocco. While Türkiye is in first place for fig exports, it is followed by the USA, Spain, Syria and Greece. Fig trees are commonly planted in gardens everywhere. They are berry trees well adapted to aridity and high temperatures in the Mediterranean region and similar climates [93].

3. Production models compatible with climate change

Now the focus is on production methods for these berry fruits that are compatible with aridity and global climate change. As is known, temperature increases will increase the degradation rate and process in soil; the resulting erosion hazard will reduce organic matter linked to humidity loss in soil. A reduction in soil productivity will be caused by these negative conditions. Innovative methods like soil-free agriculture, an increasingly common alternative production model in the agricultural field in recent years, restorative agriculture and vertical agriculture are thought to support the struggle against climate change. Of these production models, it is thought that new production models like soil-free agriculture, restorative agriculture and vertical agriculture will be appropriate for fruit cultivation, especially for some berry fruits with ground-spreading or small-bush forms. For example, in strawberry production, strawberry cultivation and traditional cultivation methods use soil; this involves several problems caused by soil-derived pathogens and nematodes causing significant product loss. To prevent these losses, alternation, solarization of soil or disinfection with chemicals is necessary. These solutions both increase production costs and damage the environment and soil. As a result, soil-free agriculture may be an important alternative solution for these problems. In addition to solving problems expected in greenhouse soils, targets like extending the production season in strawberry cultivation and yield per unit area, soil-free strawberry cultivation has become more common in commercial greenhouse production [98]. Additionally, production of fruit like blueberry, blackberry and raspberry under cover is becoming increasingly common. For example, blueberry production is very important in soil-free agriculture. The advantages can be listed as follows; pH control of the production environment, convenience for fertilizer, water, pesticides and labor, suitability for automation systems (pruning, harvesting, medicating), ease of protection from negative climate conditions, control of weeds, and ease of harvesting. Studies about breeding new varieties suitable for greenhouse production with low chilling requirements continue at full speed. As productive agricultural areas are being chosen for both residential areas and for production areas for industry, the agricultural production area is becoming smaller. As a result, producers have begun to choose intensive production methods for small areas like vertical agriculture and undercover production among agricultural production methods. In this way, producers choose new production models with the aim of offering fresh products to the market in all four seasons, to meet consumer demands and to obtain high income. With this, aim, producers in Türkiye and Far East and European countries have given up traditional production for figs and begun production with wired training systems. Here, the aim is to both obtain a very early product (in 2–3 years) and to sell fresh products to the market at high price. At the same time, these methods increase the amount of product obtained per decare or hectare by using more plants in a smaller area, as mentioned above. Thus in wired training systems, nearly 83 seedlings can be planted per decare with dense planting at 3–4 m intervals. Thus, more product is obtained from a smaller area. Though the first year installation cost is a bit higher in this type of production method, the producer has the opportunity to harvest in a shorter duration and sell to market at a higher price, increasing the profitability rate of wired training dwarf production methods. At the same time, producers choose these methods as there are lower labor costs for harvest and weed control. Again, undercover mulberry cultivation is performed with the aim of controlling climate conditions and offering fresh berry fruit to the market in a short period. The mulberry variety grown in greenhouses especially in Türkiye is the variety belonging to the *Morus laevigata* Wall. species, known as finger mulberry or black mulberry among the public. Greenhouse mulberry production in Türkiye is used in Anamur and Silifke counties in Mersin. These mulberry fruits are sold to buyers at high prices in the market as fresh fruit, not as dried mulberries. In greenhouse mulberry production, harvest begins in February and ends in June. If a single harvest is desired in open and greenhouse production in the Mediterranean region, hard pruning should be performed. If the target is a second harvest in a greenhouse, normal pruning should be performed. Additionally, the second greenhouse harvest occurs in December, January and February. As can be seen, these types of new research and new production models provide an alternative gain for producers and offer fresh fruit to consumers.

Modern gene technologies, also called biotechnology in recent years, offer important opportunities to increase agricultural production in order to provide adequate and balanced nutrition for the rapidly increasing world population. In this case, in addition to the use of sustainable farming techniques, the creation of high yielding and high quality plant types resistant to biotic and abiotic stress conditions is the main goal. In the near and medium term, it will be more correct to focus mostly on molecular plant breeding approaches, rather than simply transgenic plants obtained by transformation, in the development of these plants. Developing countries having abundant gene resources, such as Turkey, will assist them in making the greatest use of their genetic potential by identifying priority areas, providing adequate infrastructure for molecular biology studies, and training a critical mass of competent researchers. However, in parallel with technological advancements, it is necessary to make legal regulations regarding biosecurity are required, both during the creation of these techniques and products and their release into nature, as well as the training of competent personnel who will apply this legislation. In this case, the proposed legislation must be founded on scientific grounds. Likewise, it is very important for agriculturally developing countries such as Turkey to make the legislation such as Plant Breeder's Rights and Patent Law for biotechnological applications in berry fruits and intellectual property rights related to products immediately applicable, to inform and support researchers in these fields, and to make regulations and bring them to a competitive position in the globalizing world trade. The necessity of adapting to climate change is essential for berry fruits to other fruit species. Because the composition of the grape grain is a major predictor of the quality, distinctiveness, and market value of juice, frozen or dried grapes, and wine, and is strongly reliant on the natural environment or ecology, as well as the harvest season. There is a considerable desire for lowering pesticide use in berry fruits, as there is in all fruit farming around the world. There are three inseparable inputs to the equation that berry growers must solve: first, adapting to climatic change, second, minimizing pesticide use, and thirdly, preserving the aroma unique to the species and variety. Although garden management can maintain these qualities to some amount, it is believed that addressing the short-medium-term effects of climate change would require long-term sustainable solutions, rather than short-term genetic improvement. On the one hand, the positive aspects of breeding new varieties of berry fruits that adapt to climate change are as follows: (a) species and variety richness not yet fully explored; (b) advances in sequencing technologies that allow high-throughput sequencing of whole genomes, faster mapping of targeted traits, and easier identification of genetic relationships; (c) advances in new production technologies that potentially allow for definitive changes to established genes; (d) automation of phenotyping allowing faster and more complete tracking of many traits in relatively large plant populations;

(e) functional characterization of an increasing number of genes involved in developmental control, fruit metabolism, disease resistance, and adaptation to the environment. On the other hand, challenges in breeding new varieties of berry fruits that adapt to climate change are as follows: (a) The perennial nature of the garden plant and its vast size, necessitating extensive and labor-intensive field experiments; (b) low transformation and regeneration efficiency, as well as the limited size of breeding populations; (c) the complexities of transferable traits and the need to more clearly define future ideotypes; (d) a lack of shared and integrative platforms that allow full assessment of genotype–phenotype-environmental linkages; and (e) legal, market, and consumer acceptance of new genotypes [99].

The responses of berry farmers to climate change adaptation can be divided into two groups. It can be summarized as (1) short-term incremental responses (incremental responses) that farmers usually choose autonomously based on the observed changes and local knowledge and experience, and (2) long-term transformative responses that require strategic planning, usually applied on a larger spatial scale. Models can be used to aid decision making at both response levels; thus, certain model elements are either more or less helpful depending on the type of adaption response. In order to adapt to climate change, berry farmers are expected to focus on five different models: (1) empirical crop models; (2) regional conformity models; (3) biophysical models; (4) meta models; and (5) decision models. These models' potential and limitations in deciding on short- and long-term adaption planning are examined. The biggest challenge in applying these models is to consider climate variability and uncertainty, which are future research, integrated risk and vulnerability assessments that support climate change adaptation efforts. To achieve long-term compliance success, systems for monitoring management adjustments and their outcomes must be institutionalized [100].

4. Conclusion and recommendations

Climate change has intensified global food crises, while the food crises are intensified by climate change and small-scale farmers are predicted to be the group first and most severely affected by climate change. Finally all branches of the agricultural sector will be negatively affected by global warming and climate change. Here, one of the agricultural sectors most impacted by negative climate conditions are horticultural plants. Among horticultural plants, berry fruit production is an agricultural branch where production can be performed by controlling negative climate conditions. In these negative conditions, a range of new production models can be implemented to control yield loss of berry fruits and to provide high yield in a shorter duration and smaller area. For this reason, it is thought that the use of innovative production methods such as soil-free agriculture, regenerative agriculture and vertical agriculture, which are becoming increasingly common in the field of agriculture, are alternative production models that can contribute to the fight against climate change. The importance of using new agricultural models like different training systems, dwarf production allowing dense planting, undercover production, soil-free agriculture, restorative agriculture and vertical agriculture is increasing. Additionally, it is necessary to choose restricted irrigation systems like drop and leach systems for correct water management. In breeding studies for new varieties, it is very important to give weight to breeding new varieties with resistance to salinity, aridity, plant diseases and pests, good flavor and aroma, and high yield. It is necessary to implement tighter precautions to protect national and international plant gene sources paying attention to local varieties more resistant to negative natural and climate conditions in these breeding studies. It is necessary to implement multidisciplinary studies especially for berry fruit production about improving techniques for drying and processing and using new methods that are environmentally friendly and do not damage human health to ensure products reach consumers fresh and without nutritional loss or degradation. In conclusion, we believe this study will guide more detailed studies about this topic and continuation of berry fruit production with least impact from global warming and climate change conditions.

Author details

İlbilge Oğuz[1]*, Halil İbrahim Oğuz[2], Şule Hilal Attar[1], Duygu Ayvaz Sönmez[4], Hüseyin Çelik[3] and Nesibe Ebru Yaşa Kafkas[1]

1 Department of Horticulture, University of Çukurova, Adana, Türkiye

2 Department of Horticulture, Faculty of Agriculture, Adıyaman University, Adıyaman, Türkiye

3 Department of Horticulture, Faculty of Agriculture, University of 19 Mayıs, Samsun, Türkiye

4 Yaltır Agricultural Products Inc., Sarıhuğlar, Türkiye

*Address all correspondence to: ilbilge94@gmail.com

References

[1] Bustos MC, Rocha-Parra D, Sampedro I, de Pascual-Teresa S, León AE. The influence of different air-drying conditions on bioactive compounds and antioxidant activity of berries. Journal of Agricultural and Food Chemistry. 2018;**66**(11):2714-2723. DOI: 10.1021/acs.jafc.7b05395

[2] Pörtner HO, Roberts DC, Adams H, Adler C, Aldunce P, Ali E, et al. Climate Change 2022: Impacts, Adaptation and Vulnerability. Geneva, Switzerland: IPCC; 2022. p. 3056. DOI: 10.1017/9781009325844

[3] Wilhelmi OV, Wilhite DA. Assessing vulnerability to agricultural drought: A Nebraska case study. Natural Hazards. 2002;**25**:37-58. DOI: 10.1023/A:1013388814894

[4] Lesk C, Rowhani P, Ramankutty N. Influence of extreme weather disasters on global crop production. Nature. 2016;**2016**(529):84-87. DOI: 10.1038/nature16467

[5] Öztürk K. Global climatic changes and their probable effect upon Turkey. Gazi Eğitim Fakültesi Dergisi. 2002;**22**(1):47-65. Available from: https://dergipark.org.tr/en/download/article-file/77436 [Accessed: 4 March 2023] (in Turkish)

[6] Kapluhan E. Drought and drought in Turkey effect of agriculture. Marmara Coğrafya Dergisi. 2013;**27**:487-510. https://dergipark.org.tr/en/download/article-file/3335 [Accessed: 4 March 2023] (in Turkish)

[7] Kanber R, Kapur B, Ünlü M, Tekin S, Koç L. A new approach to evaluating the impact of climate change on agricultural production systems. Iccap Project. Proceeding Book of the TMMOB (Turkish Engineers and Architects) 2nd Water Policies Congress. Ankara. Türkiye. 2015. ISBN No: 978-9944-89-512-5. Available from: http://www.e-kutuphane.imo.org.tr/pdf/10917.pdf; (in Turkish) [Accessed: 23 February 2023]

[8] Varol N, Ayaz M. Global climate change and olive cultivation. Türk Bilimsel Derlemeler Dergisi. 2012;**5**(1):11-13. Available from: https://dergipark.org.tr/en/pub/derleme/issue/35085/389116 [Accessed: 4 March 2023] (in Turkish)

[9] Poyraz, Engin S, Boz Y. The latest developments at berry cultivation of Turkey. IJAAES International Journal of Anatolia Agricultural Engineering. 2019;**1**(Special Issue):108-115. Available from: https://dergipark.org.tr/en/download/article-file/904340 [Accessed: 4 March 2023] (in Turkish)

[10] Laaksonen O, Knaapila A, Niva T, Deegan KC, Sandell M. Sensory properties and consumer characteristics contributing to liking of berries. Food Quality and Preference. 2016;**53**:117-126. DOI: 10.1016/j.foodqual.2016.06.004

[11] Saavedra MJ, Borges A, Dias C, Aires A, Bennett RN, Rosa ES, et al. Antimicrobial activity of phenolics and glucosinolate hydrolysis products and their synergy with streptomycin against pathogenic bacteria. Medicinal Chemistry. 2010;**6**(3):174-183

[12] Bowen-Forbes CS, Zhang Y, Nair MG. Anthocyanin content, antioxidant, anti-inflammatory and anticancer properties of blackberry and raspberry fruits. Journal of Food Composition and Analysis. 2010;**23**(6):554-560. DOI: 10.1016/j.jfca.2009.08.012

[13] Çevik İ, Erhan M. Researches on the technological properties of some grape fruit varieties. Journal of Food and Feed Science Technology. 2014;**3**:1-6. Available from: https://dergipark.org.tr/en/download/article-file/44885 [Accessed: 4 March 2023] (in Turkish)

[14] Escodo P. Global Berry Market Trends & Opportunities. 40 Congreso Internacional de Frutos Rojos, 20-21 de Junio 2018. Huelva, España. Available from: https://congresofrutosrojos.com/wpcontent/uploads/2018/07/1.PierreEscodo.pdf: Palacio de Congresos, Casa Colón; 2018 [Accessed: 4 March 2023]

[15] Kaume L, Howard LR, Devareddy L. The blackberry fruit: A review on its composition and chemistry, metabolism and bioavailability, and health benefits. Journal of Agricultural and Food Chemistry. 2012;**60**(23):5716-5727. DOI: 10.1021/jf203318p

[16] Grace MH, Esposito D, Dunlap KL, Lila MA. Comparative analysis of phenolic content and profile, antioxidant capacity, and anti-inflammatory bioactivity in wild Alaskan and commercial Vaccinium berries. Journal of Agricultural and Food Chemistry. 2014;**62**(18):4007-4017. DOI: 10.1021/jf403810y

[17] De Souza VR, Pereira PAP, da Silva TLT, de Oliveira Lima LC, Pio R, Queiroz F. Determination of the bioactive compounds, antioxidant activity and chemical composition of Brazilian blackberry, red raspberry, strawberry, blueberry and sweet cherry fruits. Food Chemistry. 2014;**156**:362-368. DOI: 10.1016/j.foodchem.2014.01.125

[18] Diaconeasa Z, Ranga F, Rugina D, Leopold L, Pop O, Vodnar N, et al. Phenolic content and their antioxidant activity in various berries cultivated in Romania. Bulletin of University of Agricultural Sciences and Veterinary Medicine Cluj-Napoca: Horticulture. 2015;**72**:99-103. DOI: 10.15835/buasvmcn-fst:11127

[19] Çağlar M, Demirci M. Phenolic compounds in berry fruits and their importance in nutrition. European Journal of Science and Technology. 2017;**7**(11):18-26. Available from: https://dergipark.org.tr/en/download/article-file/476053 [Accessed: 4 March 2023] (in Turkish)

[20] Gündeşli MA, Korkmaz N, Okatan V. Polyphenol content and antioxidant capacity of berries: A review. International Journal of Agriculture Forestry and Life Sciences. 2019;**3**(2):350-361. Available from: https://dergipark.org.tr/en/download/article-file/874337 [Accessed: 4 March 2023] (in Turkish)

[21] Bower C, Stan S, Daeschel M, Zhao Y, State University Extension Service O. Promoting the safety of Northwest fresh and processed berries. In: Corvallis Or. USA: Extension Service, Oregon State University; 2003. pp. 1-37. Available from: https://ir.library.oregonstate.edu/concern/open_educational_resources/8s45q905t [Accessed: 4 March 2023]

[22] Skrovankova S, Sumczynski D, Mlcek J, Jurikova T, Sochor J. Bioactive compounds and antioxidant activity in different types of berries. International Journal of Molecular Sciences. 2015;**16**(10):24673-24706. DOI: 10.3390/ijms161024673

[23] Kalt W, Cassidy A, Howard LR, Krikorian R, Stull AJ, Tremblay F, et al. Recent research on the health benefits of blueberries and their anthocyanins. Advances in Nutrition. 2020;**11**(2):224-236. DOI: 10.1093/advances/nmz065

[24] Salo HM, Nguyen N, Alakärppä E, Klavins L, Hykkerud AL, Karppinen K, et al. Authentication of berries and berry-based food products. Comprehensive Reviews in Food Science and Food Safety. 2021;**20**(5):5197-5225. DOI: 10.1111/1541-4337.12811

[25] Neto CC. Cranberry and blueberry: Evidence for protective effects against cancer and vascular diseases. Molecular Nutrition and Food Research. 2007;**51**(6):652-664. DOI: 10.1002/mnfr.200600279

[26] Battino M, Beekwilder J, Denoyes-Rothan B, Laimer M, McDougall GJ, Mezzetti B. Bioactive compounds in berries relevant to human health. Nutrition Reviews. 2009;**67**(suppl_1):S145-S150. DOI: 10.1111/j.1753-4887.2009.00178.x

[27] Vendrame S, Del Bo' C, Ciappellano S, Riso P, Klimis-Zacas D. Berry fruit consumption and metabolic syndrome. Antioxidants. 2016;**5**(4):34. DOI: 10.3390/antiox5040034

[28] Jurikova T, Mlcek J, Skrovankova S, Sumczynski D, Sochor J, Hlavacova I, et al. Fruits of black chokeberry Aronia melanocarpa in the prevention of chronic diseases. Molecules. 2017;**22**(6):944. DOI: 10.3390/molecules22060944

[29] Alwazeer D. Kuru gıdaların rengini muhafaza etmeye yönelik yeni bir teknik: indirgen atmosferik kurutma. Journal of the Institute of Science and Technology. 2018;**8**(4):125-131. DOI: 10.21597/jist.418232 (in Turkish) [Accessed: 4 March 2023]

[30] Costa R, Fraga H, Fonseca A, de Cortazar-Atauri IG, Val MC, Carlos C, et al. Grapevine phenology of cv. Touriga Franca and Touriga Nacional in the DouroWine region: Modelling and climate change projections. Agronomy Basel. 2019;**9**:210. DOI: 10.3390/agronomy9040210

[31] Fraga H, Santos JA. Daily prediction of seasonal grapevine production in the Douro wine region based on favourable meteorological conditions. Australian Journal of Grape and Wine Research. 2017;**23**(2):296-304. DOI: 10.1111/ajgw.12278

[32] Şahin M, Topal E, Özsoy N, Altunoğlu E. The effects of climate change on fruit growing and beekeeping. Journal of Anatolian Natural Sciences. 2015;**6**(2):147-154 (in Turkish)

[33] Webb LB, Whetton PH, Barlow EWR. Modelled İmpact of future climate change on the phenology of wine grapesin Australia. Australian Journal of Grape and Wine Research. 2007;**13**(3):165-175. DOI: 10.1111/j.1755-0238.2007.tb00247.x

[34] Luedeling E, Girvetz EH, Semenov MA, Brown PH. Climate change affects winter chill for temperate fruit and nut trees. PLoS One. 2011;**6**(5):e20155. DOI: 10.1371/journal.pone.0020155

[35] Tasnim R, Drummond F, Zhang YJ. Climate change patterns of wild blueberry fields in Downeast, Maine over the past 40 years. Water. 2021;**13**(5):594. DOI: 10.3390/w13050594

[36] Malhotra SK, Srivastva AK. Climate smart horticulture for addressing food, nutritional security and climate challenges. In: Srivastava AK, et al, editors. Shodh Chintan- Scientific Articles. New Delhi: ASM Foundation; 2014. pp. 83-97

[37] Ruml M, Vuković A, Vujadinović M, Djurdjević V, Ranković-Vasić Z, Atanacković Z, et al. On the use of regional climate models: Implications of

climate change for viticulture in Serbia. Agricultural and Forest Meteorology. 2012;**158**:53-62. DOI: 10.1016/j.agrformet.2012.02.004

[38] Neri D, Baruzzi G, Massetani F, Faedi W. Strawberry production in forced and protected culture in Europe as a response to climate change. Canadian Journal of Plant Science. 2012;**92**(6):1021-1036. DOI: 10.4141/cjps2011-276

[39] Dal Santo S, Tornielli GB, Zenoni S, Fasoli M, Farina L, Anesi A, et al. The plasticity of the grapevine berry transcriptome. Genome Biology. 2013;**14**(6):1-18. DOI: 10.1186/gb-2013-14-6-r54

[40] Bradshaw AD. Evolutionary significance of phenotypic plasticity in plants. Advances in Genetics. 1965;**13**:115-155

[41] Fusco G, Minelli A. Phenotypic plasticity in development and evolution: Facts and concepts. Philosophical Transactions of the Royal Society B: Biological Sciences. 2010;**365**(1540):547-556. DOI: 10.1098/rstb.2009.0267

[42] Bedford T, Hartl DL. Optimization of gene expression by natural selection. Proceedings of the National Academy of Sciences. 2009;**106**(4):1133-1138. DOI: 10.1073/pnas.0812009106

[43] Ranz JM, Machado CA. Uncovering evolutionary patterns of gene expression using microarrays. Trends in Ecology & Evolution. 2006;**21**(1):29-37. DOI: 10.1016/j.tree.2005.09.002

[44] Sadras VO, Stevens RM, Pech JM, Taylor EJ, Nicholas PR, McCarthy MG. Quantifying phenotypic plasticity of berry traits using an allometric-type approach: A case study on anthocyanins and sugars in berries of cabernet sauvignon. Australian Journal of Grape and Wine Research. 2007;**13**(2):72-80. DOI: 10.1111/j.1755-0238.2007.tb00237.x

[45] Balcı G, Yakupoğlu T. Soil compaction in a small berry garden established in a semi-arid region. Turkish journal of agriculture-food. Science and Technology. 2018;**6**(10):1412-1418. (in Turkish). DOI: 10.24925/turjaf.v6i10.1412-1418.2050

[46] Gerçekcioğlu R, Ağaoğlu YS. Gooseberry and black currant. In: Ağaoğlu YS, Gerçekcioğlu R, editors. Berry Fruits. No. 1. Türkiye: Tomurcukbag Ltd. Sti. Education Publications; 2013. p. 654. (in Turkish). ISBN:978-975-978-605-64181-1-NaN

[47] Dumanoğlu H, Yeşiloğlu T, Erdoğan V, Serçe S, Demirsoy H, Akçay E, et al. Changes and New Searches in Fruit Production. Proceeding Book of the Türkiye Agricultural Engineering VIII. Technical Congress, 12 January 2014. Çankaya/Ankara (in Turkish); 2014. pp. 550-578 [Accessed: 8 March 2023]

[48] Plekhanova MN. Blue honeysuckle (*Lonicera caerulea* L.)-a new commercial berry crop for temperate climate: Genetic resources and breeding. Acta Horticulturae. Dresden (Germany). 1999;**538**:159-164. ISBN:9789066059061. DOI: 10.17660/Actahortic.2000.538.25

[49] Hummer KE, Pomper KW, Postman J, Graham CJ, Stover E, Mercure EW, et al. Emerging fruit crops. Handbook of Plant Breeding book series. Vol. 8. Boston, MA: Springer; 2012. pp. 97-147. ISBN: 978-1-4419-0763-9. DOI: 10.1007/978-1-4419-0763-9_4. Available from: https://link.springer.com/bookseries/7290

[50] Rupasinghe HV, Boehm MM, Sekhon-Loodu S, Parmar I, Bors B,

Jamieson AR. Anti-inflammatory activity of haskap cultivars is polyphenols-ependent. Biomolecules. 2015;**5**(2):1079-1098. DOI: 10.3390/biom5021079

[51] Bussieres J, Boudreau S, Clément-Mathieu G, Dansereau B, Rochefort L. Growing black chokeberry (Aronia melanocarpa) in cut-over peatlands. HortScience. 2008;**43**(2):494-499. DOI: 10.21273/HORTSCI.43.2.494

[52] Roşca I, Donea V, Donea E, Dadu C, Sava P. *Aronia melanocarpa* (Michx.) Elliot. cultură de perspectivă pentru Republica Moldova. Pomicultura, Viticultura şi Vinificația. 2019;*81*(3-4):50-57

[53] Dinu MD, Mazilu IE, Cosmulescu S. Influence of climatic factors on the phenology of chokeberry cultivars planted in the Pedoclimatic conditions of southern Romania. Sustainability. 2022;**14**(9):4991. DOI: 10.3390/su14094991

[54] Plekhanova, M.N. Blue Honeysuckle (Lonicera caerulea L.) - A New Commercial Berry Crop For Temperate Climate: Genetıc Resources and Breedıng. Acta Horticulturae. 2000

[55] Atila SP. Preliminary Evaluations on the Adaptation of Some Raspberry and Blackberry Cultivars to Ayas (Ankara) Conditions [Master thesis]. Ankara: Ankara University, Institute of Science and Technology, Department of Horticulture; 2002 [Accessed: 4 March 2023] (in Turkish)

[56] Çalışkan S. Faunistic and Systematic Studies on Plant Parasitic Nematode Species in Some Orchards Growing Raspberry (*Rubus idaeus* L.) and Blackberry (*Rubus fruticosus* L.) in Turkey [PhD thesis]. Tokat: University of Gaziosmanpaşa Graduate School of Natural and Applied Sciences, Department of Plant Protection; 2019 (in Turkish)

[57] Akkurt M. Raspberry. [PowerPoint slide]. Berry Fruits Bachelor Degrees Course Material. Ankara University. Ankara, Türkiye: Open Access System; 2021. Available from: https://acikders.ankara.edu.tr/pluginfile.php/112121/mod_resource/content/0/AHUDUDU%201.pdf [Accessed: 4 March 2023] (in Turkish)

[58] Durgo K, Belščak-Cvitanović A, Stančić A, Franekić J, Komes D. The bioactive potential of red raspberry (*Rubus idaeus* L.) leaves in exhibiting cytotoxic and Cytoprotective activity on human laryngeal carcinoma and Colon adenocarcinoma. Journal of Medicinal Food. 2012;**15**(3):258-268. DOI: 10.1089/jmf.2011.0087 [Accessed: 5 March 2023]

[59] Ferlemi AV, Lamari FN. Berry leaves: An alternative source of bioactive natural products of nutritional and medicinal value. Antioxidants (Basel). 2016;**5**(2):17. DOI: 10.3390/antiox5020017

[60] Bowman R, Taylor J, Muggleton S, Davis D. Biophysical effects, safety and efficacy of raspberry leaf use in pregnancy: A systematic integrative review. BMC Complementary Medicine and Therapies. 2021;**21**(1):56. DOI: 10.1186/s12906-021-03230-4

[61] Schulz M, Chim JF. Nutritional and bioactive value of Rubus berries. Food Bioscience. 2019;**31**:100438. DOI: 10.1016/j.fbio.2019.100438

[62] Wu L, Liu Y, Qin Y, Wang L, Wu Z. HPLC-ESI-qTOF-MS/MS characterization, antioxidant activities and inhibitory ability of digestive enzymes with molecular docking analysis of various parts of raspberry (*Rubus ideaus* L.). Antioxidants. 2019;**8**(8):274. DOI: 10.3390/antiox8080274

[63] Staszowska-Karkut M, Materska M. Phenolic composition, mineral content,

and beneficial bioactivities of leaf extracts from black currant (*Ribes nigrum* L.), raspberry (*Rubus idaeus*), and Aronia (*Aronia melanocarpa*). Nutrients. 2020;**12**:463. DOI: 10.3390/nu12020463

[64] Hummer KE. Rubus pharmacology: Antiquity to the present. HortScience. 2010;**45**(11):1587-1591. DOI: 10.21273/HORTSCI.45.11.1587

[65] Göktaþ A. Raspberry and Blackberry Cultivation. Republic of Türkiye Ministry of Agriculture and Forestry Fruit Research Institute. Ankara, Türkiye; 2011. p. 38. Available from: https://www.sorhocam. com/uploads/docs/bogurtlen-ve-ahududu-yetistiriciligi-31565.pdf [Accessed: 4 March 2023] (in Turkish)

[66] Yiğen Ç, Baki İ, Aytaç S. Organic Farming Areas and Opportunities in the Black Sea Region. Proceeding Book of the Eastern Black Sea II. Organic Agriculture Congress, 6-9 October 2015. Rize, Türkiye. 2015. pp. 318-328 [in Turkish)

[67] Karadeniz T. Some Natural Fruit Species Grown in the Black Sea Region, their Introduction and Evaluation. Organic Farming Areas and Opportunities in the Black Sea Region. Proceeding Book of the Eastern Black Sea II. Organic Agriculture Congress, 6-9 October 2015. Rize, Türkiye. 129-139 pp. [Accessed: 4 March 2023] (in Turkish)

[68] Akkurt M. Blackberry. [PowerPoint slide]. Berry Fruits Bachelor Degrees Course Material. Ankara University. Ankara, Türkiye: Open Access System; 2021b. Available from: https://acikders.ankara.edu.tr/pluginfile.php/112123/mod_resource/content/0/BÖÐÜRTLEN%201.pdf [Accessed: 4 March 2023] (in Turkish)

[69] Strik BC. A review of nitrogen nutrition of Rubus. Acta Horticulturae. 2008;**777**:403-410. DOI: 10.3390/ijms16102467310.17660/ActaHortic.2008.777.61

[70] Milošević T, Milošević N, Glišić I, Mladenović J. Fruit quality attributes of blackberry grown under limited environmental conditions. Plant, Soil and Environment. 2012;**58**:322-327. DOI: 10.1016/j.scienta.2017.09.013

[71] Clark JR, Finn CE. Blackberry cult ivation in the world. Revista Brasileira de Fruticultura. 2014;**36**:46-57. DOI: 10.1590/0100-2945-445/13

[72] Oliveira JD, Cruz MDCMD, Moreira RA, Fagundes MCP, Sena CG. Productive performance of blackberry cultivars in altitude region. Ciência Rural, Santa Maria. 2017;**47**(12):e20170021. DOI: 10.1590/0103-8478cr20170021

[73] Tanker M, Çitoğlu G, Soner Ö. Pectin Content in Rosa Fruits. Proceedings Book of the Rosehip Symposium, 5-6 September, Gümüşhane. 1996. pp. 297-307 (in Turkish)

[74] Kutbay HG, Kılınç M. Taxonomic characteristics of rosehip (Rosa L.) species and distribution in Turkey. Rosehip Symposium. 1996;**5**(6):75-83 (in Turkish)

[75] Türkben C. Rosehip. Bursa: Uludag University Press; 2003 (in Turkish)

[76] Nybom H, Rumpunen K, Erciþli S. Proceedings of the 1st International RoseHip Conference: Gümüþhane, Türkiye, September 7-10, 2004. International Society for Horticultural Science; 2005. Acta Horticulturae (690), pp

[77] Ekincialp A. Selection of Naturally Growing Rosehips (Rosa spp.) in Hakkari Center, 2007 [Master Thesis]. Yüzüncü Yıl University, Institute of Science,

Department of Horticulture; 2007. Van-2007, (in Turkish)

[78] Ağaoğlu S, Gerçekçioğlu R. Grape Fruits. Türkiye: Tomurcukbag Ltd. Sti. Education Publications No: 1; 2013. p. 654. ISBN:978-975-978-605-64181-1-NaN

[79] Erdoğan Ü, Pırlak L. Mulberry (Morus spp) production and evaluation in our country. Alatarım. 2005;**4**(2):38-43

[80] Ilgın M, Çağlar S. Mulberry Cultivation and its Importance in Kahramanmaraş. Proceedings Book of the II. National Symposium on Berries. Tokat, Türkiye. 14-16 September 2006. pp. 189-192. ISBN: 978-9944-77-142-9

[81] Freeman WH. Temprate-Zone Pomology. San Fransisco: W.H. Freeman and Company; 1978. p. 428

[82] Huo Y. Mulberry Cultivation and Utilization in China. Mulberry for Animal Production, FAO Animal Production and Healt Paper. Vol. 1472002. Rome, Italy: FAO; 2000. pp. 11-44

[83] Koidzumi G. Taxonomy and phytogeography ofthe genus Morus. Bull Seric Exp Stat Tokyo. 1917;**3**:1-62

[84] Machii H, Koyama A, Yamanouchi H, Matsumoto K, Kobayashi S, Katagiri K. A list of Motphological and agronomical traits of mulberry genetic resources. Misc. Publ. Natl. Inst. Seric. Entomol. Sci. 2001;**29**:1-307

[85] Martin G, Reyes F, Hernandez I, Milera M. Agronomic Studies with Mulberry in Cuba. Mulberry for Animal Production, FAO Animal Production and Healt Paper. Vol. 1472002. Rome, Italy: FAO; 2000. pp. 103-114

[86] Datta RK. Mulberry Cultivation and Utilization in India. Mulberry for Animal Production, FAO Animal Production and Health Paper. Vol. 1472002. Rome, Italy: FAO; 2000. pp. 45-62

[87] Dharmananda S. Fruit as Medicine, Morus Fruit (Mulberry). Portland, Oregon: Institute for Traditional Medicine; 2003. Available from: http://www.itmonline.org/arts/morus.htm

[88] Singhal BK, Khan MK, Dhar A, Baqual FN, Bindroo BB. Approaches to industrial exploitation of mulberry (mulberry sp.) fruits. Journal of Fruit and Ornamental Plant Research. 2010;**18**(1):83-99. Available from: http://www.inhort.pl/files/journal_pdf/journal_20101/full8%2020101.pdf

[89] Tanrıkulu N. Guide to Using Medicinal Plants Correctly. In: Baskı I editor. İstanbul: Hayygrup Press; 2010. 176 p. ISBN: 978-605-4325-32-0. (in Turkish)

[90] Saraçoğlu A. Herbal Cures Guide. İstanbul: Ajans Mik Press; 2009. 226p. (in Turkish)

[91] Condit IJ. Ficus: The Exotic Species. Division of Agricultural Sciences: University of California; 1969. 363 pp

[92] Ferguson L, Michailides TJ, Shorey HH. The California fig industry. Horticultural Reviews. 1990;**12**:1409-490

[93] Solomon A, Golubowicz S, Yablowicz Z, Grossman S, Bergma M, et al. Antioxidant activities and anthocyanin content of fresh fruits of common fi g (*Ficus carica* L.). Journal of Agricultural and Food Chemistry. 2006;**54**(20):7717-7723

[94] Stover E, Aradhya M, Crisosto C, Ferguson L. The fig: Overview of an ancient fruit.

HortScience. 2007;**42**(5):1083-1087. DOI: 10.21273/Hortsci.42.5.1083

[95] De Candolle A. Origin of Cultivated Plants (Reprint of 2nd Edition 1967). New York: Hafner; 1886. p. 468

[96] Crisosto H, Ferguson L, Bremer V, Stover E, Colelli G. Fig (*Ficus carica* L.). In: Postharvest Biology and Technology of Tropical and Subtropical Fruits. Sawston, UK: Woodhead Publishing Series in Food Science, Technology and Nutrition; 2011. pp. 134-158, 159e-160e. DOI: 10.1533/9780857092885.134

[97] TÜİK. 2022. Turkish Statistical Institute. https://www.tuik.gov.tr/ [Accessed: 05 March 2023]

[98] Demirsoy L, Mısır D, ve Nafiye ADAK. Strawberry production in soilless culture. Anadolu Journal of the Aegean Agricultural Research Institute (AARI). 2017;27(1):71-80. Available form: https://dergipark.org.tr/en/download/ article-file/337755 [Accessed: 05 March 2023] (in Turkish)

[99] Rugienius R, Frercks B, Mažeikienė I, Rasiukevičiūtė N, Baniulis D, Stanys V. Development of climate-resilient varieties in rosaceous berries. In: Kole C editor. Genomic Designing of Climate-Smart Fruit Crops. Cham: Springer; 2020. pp. 333-384. ISBN: 978-3-319-97946-5. DOI: 10.1007/978-3-319-97946-5_9

[100] Holzkämper A. Adapting agricultural production systems to climate change—what's the use of models? Agriculture. 2017;7(10):86. DOI: 10.3390/agriculture7100086

Chapter 2

Ecological Significance of Wild Huckleberries (*Vaccinium membranaceum*)

Janene Lichtenberg and Tabitha Graves

Abstract

Wild huckleberry (*Vaccinium globare/membranaceum* complex) is a keystone species in the Pacific Northwest of the United States. The fruits are a primary food source for grizzly bears and other wildlife, as well as an important traditional and contemporary human food. Huckleberry shrubs also provide cover and nesting habitat for many animal species, including small mammals and birds. The flowers provide nectar and pollen with crucial connections between bumble bees (*Bombus* species) and huckleberries. Native bee pollination is essential for successful berry development. Huckleberries flower early in the growing season and are some of the only floral resources available when bumble bee queens first emerge from hibernation and need to collect pollen and nectar for nesting. One of these species, the Western bumble bee (*Bombus occidentalis*), is in review for listing under the U.S. Endangered Species Act. Future climate change has the potential to influence huckleberry distribution, productivity, and phenology. These potential changes could have wide-ranging implications because of the economic, cultural, and ecological importance of huckleberry.

Keywords: *Vaccinium*, huckleberry, pollination, bumble bees, wildlife, keystone species, ecological relationships

1. Introduction

The genus *Vaccinium* (Ericales: Ericaceae) includes the domestic blueberry cultivars and occurs under a variety of locations and conditions worldwide. There are more than 35 species native to North America alone, including a variety of species referred to as wild huckleberry in the Pacific Northwest. There are at least six species of *Vaccinium* in Montana, USA [1]. This chapter will focus on the huckleberry (*V. membranaceum* Douglas ex Torrey) also referred to in some sources as (*V. membranaceum/globare* complex), two separated species that include (*Vaccinium globare* Ryberg) and *V. globare* as a variety of *V. membranaceum* [2]. There are also a wide variety of common names used for this plant including wild, thinleaf, black, blue, mountain, common, big, and square-twigged huckleberry. This species is also the most common western huckleberry harvested for commercial use [3]. It is distributed from the Northwest Coast to eastern Wyoming, USA, and north to British Columbia and

Alberta, Canada. We will refer to this species (or species complex) as "wild huckleberry" or simply huckleberry in this chapter (**Figure 1**).

The shrubs of wild huckleberry are deciduous and grow up to 2 m tall, often in rhizomatous clumps [1, 4, 5]. Leaves are oval to elliptical, 11–50 mm, usually green, and finely serrated. Flower petals are fused, giving them a globe-like shape, and are light pink or sometimes white. They grow individually on the underside of stem nodes. The berries are red to deep purple when ripe with many small (1 mm) seeds. Roots may be as deep as 100 cm with rhizomes usually occurring at depths of 8–30 cm. Huckleberries reproduce through both seeds and rhizomes with patches often consisting of vegetative clones [1, 4, 5].

Wild huckleberries are usually in areas with low canopy cover, fine and well-drained loam soils, moderate slopes, and winter precipitation as snow [5, 6]. Periodic fire (every 20–50 years) is important to prevent competition from other plants and to maintain relatively open overhead canopy although timber harvesting, avalanche chutes, and even ski runs can also lead to suitable conditions [7, 8].

Wild huckleberry shrubs are commonly associated with beargrass (*Xerophyllum tenax*) [9]. Depending on location and environmental conditions, plants may be in forests dominated by Engelmann spruce (*Picea engelmannii*), western larch (*Larix occidentalis*), limber pine (*Pinus flexilis*), ponderosa pine (*P. ponderosa*), lodgepole pine (*P. contorta*), western white pine (*P. monticola*), hemlock (*Tsuga* sp.), fir (*Abies* sp.), Douglas fir (*Pseudotsuga menzeseii*), and western red cedar (*Thuja plicata*) [5]. Other shrubs that are often found in the understory are currant (*Ribes* sp.), snowberry (*Symphoricarpos* sp.), false huckleberry (*Menziesia ferruginea*), bearberry (*Arctostaphylos uva-ursi*), white

Figure 1.
Huckleberry phenophases (life stages) showing flower buds (top left), flowers (top middle), developing green berries (top right), ripe berries (bottom left), ripe berries and red leaves (bottom middle), and harvested berries (bottom right) (SKC, USGS photos).

spirea (*Spirea betulifolia*), mountain heath (*Phyllodoce empetriformis*), Cascade azalea (*Rhododendron albiflorum*), mountain-ash (*Sorbus sitchensis*), raspberry (*Rubus* sp.), Rocky Mountain maple (*Acer glabrum*), dogwood (*Cornus* sp.), Oregon grape (*Mahonia repens*), grouse whortleberry (*Vaccinium scoparium*), bilberry (*Vaccinium deliciosum*), and other huckleberry species (*Vaccinium* sp.) [5].

Wild huckleberry is one of the most important food plants in the Pacific Northwest of North America [10, 11] and has strong cultural [10–13], economical [10, 14, 15], and ecological value [2, 16]. The berries of huckleberries are highly valued by the people of the Pacific Northwest, past and present. They are a cultural keystone species for many Native American tribes that continue traditions and ceremonies surrounding collection, storage, and consumption of berries [10–13], as well as caretaking of huckleberry patches by spreading berries and seeds, pruning berry bushes, and setting fires to clear dense vegetation and promote growth [2, 10–13]. Fruits are often eaten fresh but also cooked or dried [4]. Some tribes use the berries in pemmican or sun cakes, and to make lavender dye [4]. Various parts of huckleberry plants are medicinal. Some people use the leaves for tea or mix them with other plants for smoking [10].

Early frontier people utilized huckleberries as subsistence food [15]. Today, harvesting huckleberries is an activity many people enjoy as a special treat. People mostly eat fresh berries or bake them in foods, as well as preserve them canned, frozen, and as jams and jellies. Huckleberries are difficult to cultivate, and so most commercial berry products are from wild harvested berries [3], with most companies paying pickers to harvest berries for their products [7, 8]. Some commercial pickers harvest huckleberries to supplement other income and others sell huckleberries to meet financial needs [15].

Huckleberry shrubs provide cover, nesting habitat, and food for many animal species, including small mammals and birds [17]. The flowers of *Vaccinium* species provide nectar and pollen for animals, including at least 41 native bee species in Montana [18], and are used especially by bumble bees [18, 19]. The berries are the main source of food for wildlife [17] although some herbivores such as moose (*Alces alces*) [20], elk (*Cervus canadensis*) [21], and mule deer (*Odocoileus hemionus*) [22] also eat the stems and leaves. A significant portion of the diet of grizzly bears (*Ursus arctos*) and black bears (*Ursus americanus*) is obtained from huckleberries [8, 16, 23–26]. In Glacier National Park, Montana, USA, huckleberries can make up over 15% of a bear's diet by volume [26]. Huckleberries have some of the highest apparent digestible energy of plants consumed by bears [23]. Berry productivity can influence bear movements, emigration, and recruitment [8, 27].

Meteorological conditions often determine annual huckleberry phenology and distribution [28]. Huckleberry productivity in the Cabinet-Yaak Region was positively correlated with cooler, wetter springs, and 70% of the inter-annual variability was linked to April–June growing degree days and July temperature range based on a twenty-one year dataset [16]. Livestock grazing, mining, and the introduction of invasive weed species are often detrimental [12]. Fire suppression creates unfavorable conditions for huckleberry plants due to denser forest canopy and more competition from understory plants. Fire was an important tool for the Salish and Pend d'Oreille people who burned low-intensity fires to remove unhealthy plants and simulated higher berry productivity [13]. Much of the United States and Canada began preventing fire setting by indigenous people as early as the late 1800s [10]. Although there is growing recognition by mainstream forest managers that prescribed fire is important for the health of montane ecosystems, fuel buildup from past fire suppression has led to implementation challenges [29].

Furthermore, the Pacific Northwest is experiencing warming temperatures (0.7 degrees C warmer on average) and less precipitation [30], leading to decreases in snowpack and soil moisture. These factors are shifting the timing of berry ripening in some areas; thus, timing for harvesting plants has become less predictable [28, 31]. Models are being used to help determine how climate change might influence huckleberry distribution and phenological stages (flowering, berry ripening) in the Pacific Northwest region [28]. A study investigating how climate change could affect the wild huckleberry concluded that the suitable range of huckleberry across the Pacific Northwest United States will likely shrink by 5–40% by the end of the twenty-first century [28]. The study further suggests that the phenology of huckleberry will change, with flowering advancing 23–50 days and fruiting advancing 24–52 days, by the end of the twenty-first century [28]. Models are also being developed to predict sites where huckleberries are most likely to be found for human harvesting and bears [6, 32–34].

Climate fluctuations leading to changes in snow depth and timing of snowmelt may lead to phenological mismatches, thus interrupting species interactions and reducing populations of some species. Earlier snowmelt and shifts in the timing of flowering can increase exposure of flowers to frost and reduce berry production, resulting in years with poor to no berries [35–38]. Furthermore, frost damage to flowers can impact pollinators. For example, fecundity of the alpine butterfly (*Speyeria mormonia*) was reduced due to frost damaging the flowers of their preferred floral resource (*Erigeron speciosus*) [39]. Lack of insulating snow and unpredictable and repeated spring freeze–thaw cycles may damage some insect populations and ranges [40]. Foothills and valley bottoms of the Southern Canadian Rocky Mountains are projected to lose 50% of their species richness by 2080 [33] as a result of upslope migrations of plants.

There are also concerns that climate change could increase population size and spread of pest species and pathogens. Warmer temperatures are expected to make conditions more suitable for invasive plant species, resulting in increases at higher elevations and latitudes [41]. Invasive plants tend to have rapid growth rates, wide dispersal, and the ability to outcompete native plants [41]. Many of the pest insect species in the Order Hemiptera (true bugs) have piercing/sucking mouthparts and can cause economic damage [42]. Three hemipteran families collected from the leaves and stems of wild huckleberry plants in Montana, Cicadellidae (leafhoppers), Aphidadae (aphids), and Miridae (plant bugs) include species known to cause damage to fruit crops and two species in particular, the leafhopper (*Dikraneura carneola*) and the plant bug (*Lygus hesperus*), are of particular concern [18]. A species in the Order Diptera, Spotted wing drosophila (*Drosophila suzukii*) is a fruit fly from Asia invasive to North America and may increase, and potentially impact huckleberry plants, as the climate becomes warmer and more humid [43]. Soil moisture deficits from warm and dry conditions can contribute to Western hemlock looper larvae (*Lambdina fiscellaria lugubrosa* Hulst) outbreaks that can impact not only western hemlock (*Tsuga heterophylla*) but other forest species as well including deciduous species [44] such as huckleberries. We observed caterpillars of this and other species eating huckleberry leaves and documented in a few isolated cases the consequent death of the shrub.

Species-specific responses to climate change can also disrupt plant-pollinator interactions [45, 46]. Environmental cues such as air temperature and day length influence the phenological responses of plants, insects, and other animal species [35, 40, 47, 48]. Snowpack can play a major role in the emergence of plants and insects [49–51]. In Japan, early snowmelt also increased the risk of phenological mismatch between a spring ephemeral plant and its bumble bee pollinators, likely because snow melt happened sooner than increasing soil temperatures [52]. The indirect effects on flowers can even influence bee abundances [53].

Pollinator declines and shifts in assemblages are already impacting wild plant species in some places [54]. Most commercially grown *Vaccinium* (domestic blueberries) rely on insect pollination [55], and climate change and other factors are resulting in pollinator declines [56–59]. Insect pollination is known to increase fruit or seed quality and quantity for 68% of the major crops worldwide [60], and the majority of wild plant species are dependent on insect pollination for fruit and seed set [61].

Greater than 85% of fruit set and yield of plants in the genus *Vaccinium* is dependent on pollination by insects [62], and the most efficient pollinators of many *Vaccinium* sp. are bumble bees [63, 64]. Fruit set, seed number, and fruit weight of lowbush blueberry (*V. angustifolia* and *V. myrtilloides*) was positively correlated with bee density and thus pollination [62]. Fruit development in highbush blueberry (*V. corymbosum*) was also positively correlated with insect pollination, particularly by bees [60, 63–66]. Bumble bees also increase the mass and yield of domestic cranberries (*V. macrocarpon*) [64]. Pollination success is not only linked to bee abundance but also species richness [66].

Potential changes in the habitat, health, productivity, abundance, and distribution of huckleberry plants and the species they interact with have wide-ranging implications because of the economic, cultural, and ecological importance of huckleberry. Climate change has a high potential to impact huckleberry ecology and the species they interact with such as pollinators. To better understand the potential impact of climate change on wild huckleberry, we conducted an array of studies on huckleberry phenology and plant visitors (particularly pollinators) during the years of 2015 through 2021.

2. Research on the Flathead Indian reservation, Montana

2.1 Study area

The research described in this section took place on the Flathead Indian Reservation of the Confederated Salish and Kootenai Tribes in northwest Montana, USA (**Figure 2**). Ten study sites were selected based on aspect, elevation (1200 to 1900 m), slope, plant community, geographical location, and accessibility. Dominant tree species recorded at these sites were Douglas fir, ponderosa pine, western larch, lodgepole, subalpine fir (*Abies lasiocarpa*), and grand fir (*A. grandis*). Common shrubs were snowberry, grouse whortleberry, false huckleberry, alder (*Alnus viridis*), currant, Rocky Mountain maple, serviceberry (*Amelanchier alnifolia*), wild rose (*Rosa sp*), ocean spray (*Holodiscus discolor*), spiraea, and Oregon grape. Beargrass was common at some, but not all, of the sites.

The average annual temperature in northwest Montana is 4.8°C, ranging from −4.6°C in winter to 14.7°C in summer [67]. Temperatures are cooler and more variable at higher elevations. The state growing season averaged 206 days (1981–2010). The average annual precipitation in northwest Montana is 82.2 cm with a range of an average winter minimum of 23.9 cm to an average summer maximum of 15.5 cm. Higher mountain elevations can receive as much as 150.8 cm annually. Most winter precipitation in Montana is snow that accumulates as snowpack [67].

Based on data from the nearest USDA NRCS SNOTEL station within the elevation range at our sites (Bisson Creek, 1500 m), snow was present during the months of

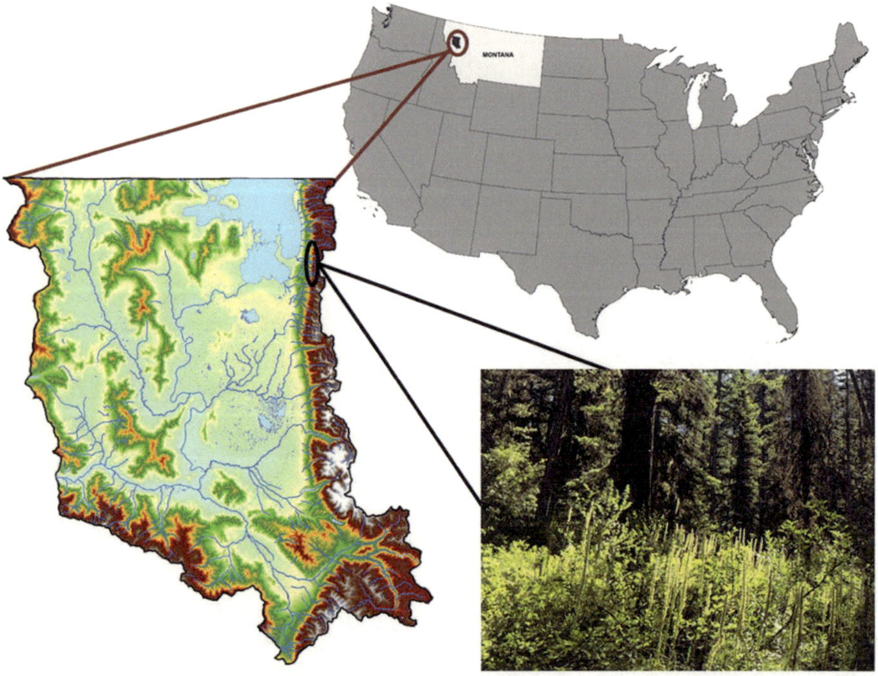

Figure 2.
Location of the Flathead reservation in northwestern Montana, USA, showing topography and landmarks within the reservation boundary, and an example of one of the study sites with huckleberry and beargrass in the foreground (CSKT maps, SKC photo).

December through April for the winters of 2014–2015 and 2015–2016 [68]. During the winters of 2016 through 2021 snow was similarly present in December but also continued into the month of May. During this study, the highest snow water equivalent (SWE) recorded was 348 mm at the start of April 2018 and the lowest SWE recorded in April was 124 mm in the year 2015. Mean spring precipitation (March–June) was the highest in 2017 (666 mm) followed by 2018 (590.25 mm) and 2015 (548.5 mm). Spring precipitation was the lowest in 2021 (419.75 mm) followed by 2020 (421.25 mm), 2016 (472.5 mm), and then 2019 (485.75 mm) [68].

2.2 Huckleberry and vertebrate visitors

A variety of methods were utilized to record huckleberry visitors including standardized bird point counts, standardized insect observations, incidental sightings, and remote cameras. During 12 hours of point count surveys at 4 sites (each visited 3 times), we documented a total of 18 species and 388 individual birds. Of these 18 species, those known to eat or that were observed eating ripe huckleberries were ruffed grouse (*Bonasa umbellus*), cedar waxwing (*Bombycilla cedrorum*), American robin (*Turdus migratorius*), Canada jay (*Perisoreus canadensis*), and Western tanager (*Piranga ludoviciana*).

Bird species photographed in the huckleberry plants while berries were ripe included ruffed grouse, American Robin, Canada jay, Western tanager, and chipping sparrow (*Spizella pallida*). Dark-eyed junco (*Junco hyemalis*), spotted towhee (*Pipilo maculatus*), song sparrow (*Melospiza melodia*), MacGillivray's warbler (*Geothlypis*

Figure 3.
Some of the bird species photographed near huckleberry plants at study sites located on the Flathead reservation. Calliope hummingbird (top left), calliope hummingbird before daylight (top right), ruffed grouse (bottom left), and dark-eyed junco (bottom right) (SKC, USGS camera photos).

tolmiei), pine grosbeak (*Pinicola enucleator*), black-capped chickadee, and great horned owl (*Bubo virginianus*) were also photographed on or near huckleberry plants outside of the ripe berry season. There were multiple sightings of calliope hummingbird (*Selasphorus calliope*) visiting huckleberry flowers during peak flowering in person and in photographs including during a rain event and before daylight (**Figure 3**).

Mammals photographed feeding on huckleberries were chipmunk (*Neotamias sp*), deer mouse (*Peromyscus maniculatus*), red squirrel (*Tamiasciurus hudsonicus*), and black bear. White-tailed deer (*Odocoileus virginianus*), elk, snowshoe hare (*Lepus americanus*), and Northern flying squirrel (*Glaucomys sabrinus*) were photographed near huckleberry plants, but we did not have any pictures of feeding (**Figure 4**). We also observed moose scat and tracks close to huckleberry plants at one study site.

2.3 Huckleberry flowers and insect visitors

Flower-visiting insects were more specifically targeted with time-constrained surveys, net captures, photographs, and incidental sightings. The majority of visitors observed were in the Order Hymenoptera (bees, wasps, hornets, ants). Some Lepidoptera (butterflies and moths) and Diptera (flies) were observed including several bee mimics; clear-wing moths (*Hemaris* sp.), day sphinx moths (*Proserpinus* sp.), and hoverflies (Family Syrphidae). Other insect visitors included beetles, ladybugs, and aphids.

Standardized, time-constrained (15 min) focal surveys of individual huckleberry plants were used to quantify the percentage of insects visiting huckleberry flowers. We conducted 359 focal surveys (approximately 90 hours of observations) during

Figure 4.
Some of the mammal species photographed near huckleberry plants at study sites located on the Flathead reservation. Snowshoe hare (top left), northern flying squirrel (middle left), white-tailed deer fawn (bottom left), black bear (top right), and deer mouse (bottom right) (SKC, USGS camera photos).

huckleberry flowering 2018–2020. Insect visitors included 569 individuals observed during 309 of the surveys. The group most frequently visiting flowers was bees (50.4%), followed by vespids (wasps and hornets, 18.3%), dipterans (flies, 18.1%), large ants (6.5%), and other [lepidopterans (moths and butterflies, 2.5%), coleopterans (beetles, 1.8%), and unidentified/other, 2.1%] (**Figure 5**). On multiple occasions, vespids were observed chewing through flower petals. Small ants were not counted because when observed there were usually large numbers moving around the plant. Bumble bees made up 85.6% of the total bees observed visiting flowers, the majority of bumble bees appearing to be queens based on size, which also aligns with the timing of queen bumble bee activity.

Most flower visitors other than bees do not actively forage for, or collect, pollen as a source of food for developing larvae. Vespids typically visit flowers either for nectar or to find prey [69] and our observations of vespids chewing holes in flowers are a sign of nectar robbing [70]. Ants feed on nectar and may move pollen but they probably have a negligible role in pollination [71]. Some dipterans may feed on nectar, and some may be pollinators [72]. Lepidopterans are pollinators but were only occasionally observed visiting huckleberry flowers. Because bees were the most common huckleberry flower visitors and are known to be efficient pollinators, and many species specifically collect pollen as a food source for developing larvae; we concluded that bees, especially bumble bees, are the main pollinators of huckleberry flowers.

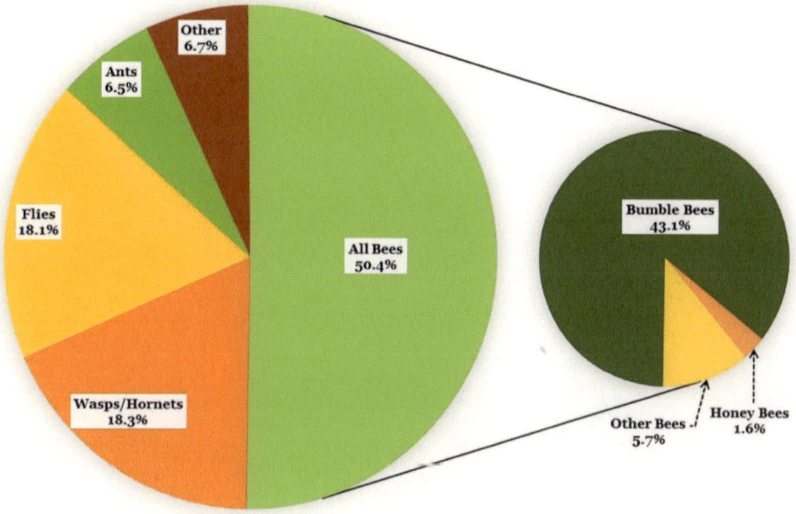

Figure 5.
Insects observed visiting huckleberry flowers at study sites on the Flathead reservation, Montana, USA.

2.4 Bumble bees as primary probable pollinators

Bumble bees are capable of sonication (buzz pollination) which might be necessary to rapidly vibrate the globe-like flowers to access pollen from the recessed anthers [63, 64, 73]. Some other native bee species are also capable of sonication [19, 73] but only contributed to a small proportion of the total flower visitors. European honey bees can be efficient pollinators, but only nine individuals were observed during focal surveys at a single site and they are not capable of sonication [73]. Honey bees likely came from hives cultivated for honey at a lower elevation. This species is not common in huckleberry habitats in Montana where most cultivated hives are not located near montane forested areas.

Because we observed bumble bees to be the most frequent visitor of huckleberry flowers, we specifically captured them for non-lethal species identification. Bumble bees were captured with sweep nets at the huckleberry sites between the years 2018 and 2021 using 45-minute standardized surveys. Each captured bee was placed in a vial labeled with the flower they were observed visiting if applicable then chilled in a cooler. Once the 45-minute capture period was complete, chilled bees were placed on gray cardboard, photographed from multiple angles using a close-up lens with ring lighting, then released. The photographs were submitted to a bumble bee specialist for identification.

There were 192 bumble bees that were both observed visiting flowering huckleberry plants and identifiable to species with confidence. From these, we identified 10 bumble bee species visiting huckleberry flowers. These species were *Bombus mixtus* (fuzzy-horned bumble bee), *B. vancouverensis,* formerly *B. bifarius* (two-formed bumble bee), *B. melanopygus* (black-tailed bumble bee), *B. sitkensis* (Sitka bumble bee), *B. flavifrons* (yellow head bumble bee), *B. rufocinctus* (red-belted bumble bee), *B. centralis* (central bumble bee), and *B. occidentalis* (Western bumble bee); as well as

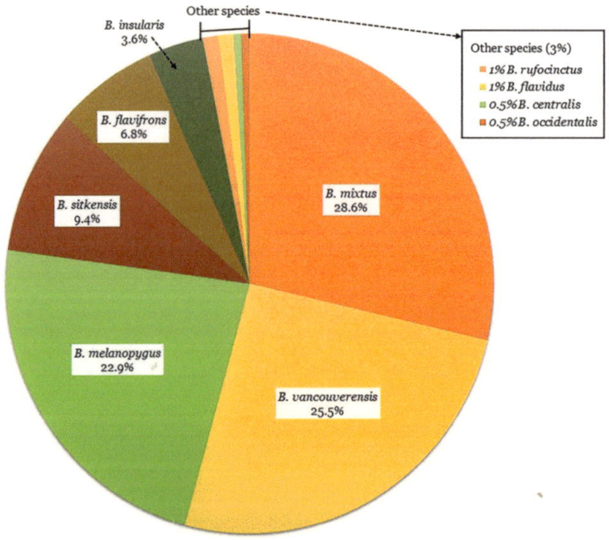

Figure 6.
Bumble bees captured from flowering huckleberry plants on the Flathead reservation, Montana, USA, and identified to species from photographs.

two cuckoo bumble bee species *B. insularis* (indiscriminate cuckoo bumble bee) and *B. flavidus* (Fernald cuckoo bumble bee) (**Figure 6**). The cuckoo bumble bees are nest parasites in the subgenus *Psithyrus* and as such do not actively collect and carry pollen. A few individuals of half-black bumble bee (*B. vagans*), Hunt's bumble bee (*Bombus huntii*), and Nevada bumble bee (*B. nevadensis*), were also captured at huckleberry sites but not from flowering huckleberry plants. Similar to prior work in Montana [18], we found the same four bumble bee species accounted for the majority (86.4%) of the total captured and identified: *B. mixtus, B. vancouverensis, B. melanopygus*, and *B. sitkensis* (**Figure 6**).

2.5 Huckleberries need bumble bee pollination

To determine the value of pollination for berry production, field experiments are needed and yet uncommon [74], particularly studies that consider the influence of pollination of wild plants using exclusion experiments [75]. We conducted a pollinator exclusion field experiment to quantify the potential importance of pollination for the development of huckleberries (**Figure 7**). We selected 120 huckleberry plants at one of the lower elevation (1250 m) study sites near Polson, Montana, USA, for this work. Half of these plants were covered with mesh bags sewn from nylon tulle netting to exclude natural pollinators and half were artificially hand-pollinated using pollen wands resulting in four treatments; (1) extra pollination: natural pollinators able to access flowers and plants also artificially pollinated by hand, (2) natural pollination: pollinators able to access flower but no additional pollination, (3) artificial pollination: natural pollinators excluded but plants artificially pollinated by hand, and (4) no pollination: natural pollinators excluded and no additional pollination. Pollination treatments (extra pollination [630 berries], natural pollination [545], and artificial pollination [365]) had far more berries than the no pollination plants (63 berries).

Figure 7.
Photograph of a subset of the 120 huckleberry plants (60 within tulle netting) included in a pollinator exclusion study on the Flathead reservation, Montana, USA (SKC photos).

Based on our observations, standardized focal surveys, and targeted captures, bumble bees were the primary visitors to huckleberry flowers. When bumble bees and other insects were excluded from pollinating huckleberry flowers, berry development was greatly limited. Even in this highly natural system at a site with overall high productivity in the year of our research, we saw a significant effect of pollen limitation on berry production.

3. Conclusions

Global climate change is resulting in warming temperatures and less precipitation in the Pacific Northwest [30]. Montana is also experiencing warmer temperatures that are predicted to increase by 3.1 to 5.4 degrees C by the end-of-the-century [67]. Precipitation in Montana has shown the greatest declines during winter months, particularly in northwest Montana. In Montana's mountains, 62–65% of annual precipitation comes from snow. Snowpack in Montana has declined since the 1930s, particularly in the mountains surrounding the Continental Divide, and snowmelt is occurring earlier in the spring. Both of these trends are likely to continue with warming temperatures [67] and have led to large shifts in the phenological start of spring for vegetation, though the direction of change varies spatially [76]. In addition, forest pests and pathogens are predicted to increase [67]. Tribal elder observations further support shifts in weather patterns, less snow, and warmer temperatures in Montana, from observations within their lifetimes and knowledge gained from prior generations [77].

Climate-related changes have resulted in documented shifts in the range of huckleberry plants and timing of berry ripening in some locations [28, 33]. Models further suggest that the range of huckleberry will decrease at lower altitudes and at the southern edge of their distribution range and perhaps shift to higher altitudes and northward, although germination or other constraints may influence this [28, 33]. Such changes in location, number, health, and timing for accessing traditional foods may alter cultural connections and is a concern of tribal communities [77]. The Séliš and Q̓ʼlispė people refer to August as Sʼtšá Spq̓ni (Month of the Huckleberry) yet at some sites we found ripe berries as early as June. Tribal elder interviews concerning climate change mention huckleberries specifically. For example, Sadie Saloway said, "The huckleberries are just disappearing so fast" and Stephen SmallSalmon said, "We changed that...about the weather. You could tell by the choke cherries getting smaller and the huckleberries getting smaller. I saw that happen." Threats posed by climate change have the potential to threaten native cultures and traditional ways of life [77].

The potential shifts in huckleberry range, numbers, and phenology could also impact the crucial interactions between bumble bees and huckleberries. In Colorado, there is documentation of a growing phenological mismatch between glacier lily (*Erythronium grandiflorum*) and the emergence of its primary pollinator, the Western bumble bee [78], a species under consideration for listing at the time of publication under the U.S. Endangered Species Act [79] and also found visiting huckleberry flowers at our field sites and others in Montana [18]. Similar mismatches in phenology could potentially become an issue with huckleberry and the bumble bee species necessary for pollination if changing conditions lead to differences in flowering and the spring emergence of queen bumble bees. Increased variability in weather in some years that leads to long periods of rain, high winds, or cold temperatures that limit the ability of bees to fly around during flowering could also restrict pollination. Based on our research, mismatched phenology or other climate-related changes could result in years with few to no berries available for the multiple species that depend on them.

Changes in huckleberry availability and phenology could also have concerning impacts for bumble bees in montane forest habitats. Huckleberries are some of the first and only flowering plants available in montane forest sites in northwest Montana when bumble bee queens emerge from hibernation in the spring. Recent research highlights the role of forests during the queen lifestage [80]. Bumble bees at these sites likely rely on early-flowering plants such as huckleberry for nectar and pollen to begin new colonies. Seed production can be reduced when flowering of spring ephemeral plants proceeds the emergence of bumble bee pollinators [46].

Declining numbers of bumble bees could also be of concern to huckleberry ecology. It is estimated that as many as one in four North American bumble bee species are being negatively impacted by some level of stress [81]. Habitat changes may be the primary stressor threatening bumble bee populations as well as stress from parasites, pesticides, and lack of flowers [82, 83] and climate change is predicted to lead to further bumble bee declines and range losses in the future [84, 85].

This chapter has illustrated the exceptional value of wild huckleberry to the ecology and people of the Pacific Northwest of North America. Our work in Montana has further added documentation of the birds, mammals, and insects that are at least incidental visitors to huckleberry plants. We added to the knowledge of species that specifically visit huckleberry flowers for pollen and nectar. We were able to provide further support that the primary probable pollinators of huckleberry flowers are bumble bees. Finally, our research used exclusion experiments in the wild to definitively quantify the significance of bumble bee pollination for the development of huckleberries. More information is needed to better understand the ideal conditions for huckleberry health and productivity, ability to spread to newly suitable areas, potential impacts of fire suppression and climate change, ecological role of interacting species, potential phenological mismatches, and conservation actions to ensure healthy huckleberry harvests into the future.

Acknowledgements

We are grateful to the Confederated Salish and Kootenai Tribes of the Flathead Reservation for permissions to conduct this work on tribal lands. We would like to thank all of the people that contributed to these projects. Dr. Laura Burkle and Casey Delphia provided feedback on our initial pollinator exclusion design and Amy Dolan for identifying bumble bees from field photos. Many people helped with the set-up, data collection, and lab work related to the exclusion study including Morgan Schultz, Madison Dimacle, Marcy Mead, Dennis Lichtenberg, Anthony Lichtenberg, Dion Lichtenberg, and Brett Stevenson. In the following years, Rebekah Brassfield served as a crew leader for an array of student research internship projects that included maintaining remote field cameras and data loggers as well as additional bumble bee focal studies and captures. Additional students that contributed to this project were Amanda Spencer, Rustin Bielski, Aspen Jaeger, Kendra Melanson, and Sarah Zurkee. Much of this work was funded by the National Science Foundation (NSF) Tribal College and University Small Grant for Research award number NSF TCP 1721243. Additional funding sources included the NSF Research Experience for Undergraduates Sustainable Land and Water Resources award number NSF EAR 1757451. The U.S. Geological Survey Species Management Program supported Tabitha Graves. Any use of trade, firm, or product names is for descriptive purposes only and does not imply endorsement by the U.S. Government.

Author details

Janene Lichtenberg[1*] and Tabitha Graves[2]

1 Wildlife and Fisheries Department, Salish Kootenai College, Pablo, USA

2 U.S. Geological Survey, Northern Rocky Mountain Science Center, West Glacier, USA

*Address all correspondence to: janene_lichtenberg@skc.edu

References

[1] Montana Natural Heritage Program. Montana field guide. 2022. Available from: https://fieldguide.mt.gov/displaySpecies.aspx?family=Ericaceae [Accessed November 11, 2022]

[2] Keefer M, Cocksedge W, Munro R, Meuleman J, MacPherson N. What about the Berries? Managing for Understory Species. British Columbia: The Centre for Livelihoods and Ecology Royal Roads University; 2010. p. 89

[3] Vander Kloet SP. *Vaccinium membranaceum*. Flora of North America Association [Internet]. 2020. Available from: https://floranorthamerica.org/Vaccinium_membranaceum [Accessed November 26, 2022]

[4] USDA (United States Department of Agriculture). NRCS (Natural Resources Conservation Service), PLANTS Database [Internet]. 2022. Available from: https://plants.usda.gov/home [Accessed November 26, 2022]

[5] USDA Forest Service. Fire Effects Information System (FEIS), Species: Vaccinium membranaceum [Internet]. 2022. Accessible from: https://www.fs.usda.gov/database/feis/plants/shrub/vacmem/all.html#27 [Accessed November 26, 2022]

[6] Proctor M, Lamb C, MacHutchon G. The Grizzly Dance of Berries and Bullets: The Relationship between Bottom up Food Resources, Huckleberries, and Top Down Mortality Risk on Grizzly Bear Population Processes in Southeast British Columbia. Kaslo (British Columbia): Trans-border Grizzly Bear Project; 2017

[7] Horowitz E. Huckleberry hounds: Sniffing out Montana's perfect gem. Montana Outdoors. 2004

[8] Martin P. Factors influencing globe huckleberry fruit production in Northwestern Montana. Bears: Their Biology and Management. 1980;**5**:159-165

[9] Simonin KA. Fire effects information system: *Vaccinium membranaceum.* [Internet]. 2018. Available from: https://www.fs.fed.us/database/feis/plants/shurb/vacmem/all.html. [Accessed May 29, 2018]

[10] Richards RT, Alexander SJ. 2A Social History of Wild Huckleberry Harvesting in the Pacific Northwest. Portland (Oregon): Pacific Northwest Research Station; 2006. p. 113. DOI:10.2737/PNW-GTR-657

[11] Trusler S, Johnson LM. Berry patch as a kind of place- the ethnoecology of black huckleberry in Northwest Canada. Human Ecology. 2008;**36**:553-568. DOI:10.1007/s10745-008-9176-3

[12] Turner NJ, Douglas D, Mellott CR. Up on the mountain: Ethnobotanical importance of montane sites in the Pacific coastal North America. Journal of Ethnobiology. 2011;**31**(1):4-43. DOI:10.2993/8-0771-31.1.4

[13] Salish C, Tribes K. Fire on the Land: An Interactive DVD [DVD]. Omaha (NB): University of Nebraska Press; 2005

[14] Thomas MG, Schumann DR. Chapter 2. Berries and wild fruit. In: Thomas MG. Schumann DR. Business Opportunities in Special Forest Products: Self-Help Suggestions for Rural Entrepreneurs, Agriculture Information Bulletin AIB-666. Lincoln (NB): Nebraska Forest Service; 1993. pp. 17-24. https://digitalcommons.unl.edu/nebforestpubs/46

[15] Carroll MS, Blatner KA, Cohn PJ. Somewhere between: Social embeddedness and the spectrum of wild edible huckleberry harvest and use. Rural Sociology. 2010;**68**(3):319-342. DOI:10.1111/j.1549-0831.2003.tb00140.x

[16] Holden ZA, Kasworm WF, Servheen C, Hahn B, Dobrowski S. Sensitivity of berry productivity to climatic variation in the Cabinet–Yaak grizzly bear recovery zone, Northwest United States, 1989-2010. Wildlife Society Bulletin. 2012;**36**(2):226-231. DOI:10.1002/wsb.128

[17] Martin AC, Zim HS, Nelson AL. American Wildlife and Plants: A Guide to Wildlife Food Habits. New York: Dover Publications; 1980. p. 2051

[18] Dolan AC. Insects Associated with Montana's Huckleberry (Ericaceae: *Vaccinium globulare*) Plants and the Bumble Bees (Hymenoptera: Apidae) of Montana. Bozeman: Montana State University; 2016

[19] Cane JH, Payne JA. Regional, annual, and seasonal variation in pollinator guilds: Intrinsic traits of bees (Hymenoptera: apoidea) underlie their patterns of abundance at Vaccinium ashei (Ericaceae). Annals of the Entomological Society of America. 1993;**86**:577-588. DOI:10.1093/aesa/86.5.577

[20] Pierce JD. Shiras moose forage selection in relation to browse availability in north-Central Idaho. Canadian Journal of Zoology. 1984;**62**:2404-2409. DOI:10.1139/z84-354

[21] Edge WD, Marcum CI, Olson-Edge SI. Summer forage and feeding site selection by elk. Journal of Wildlife Management. 1988;**52**(4):573-577. DOI:10.2307/3800909

[22] Klebenow DA. A montane forest winter deer habitat in western Montana. Journal of Wildlife Management. 1985;**29**(1):27-33. DOI:10.2307/3798627

[23] McLellen BN, Hovey FW. The diet of grizzly bears in the Flathead River drainage of Southern British Columbia. Canadian Journal of Zoology. 1995;**73**(4):704-712. DOI:10.1139/z95-082

[24] Welch CA, Keay J, Kendall KC, Robbins CT. Constraints on frugivory by bears. Ecology. 1997;**78**:1105-1119. DOI:10.2307/2265862

[25] McLellan BN. Some mechanisms underlying variation in vital rates of grizzly bears on a multiple use landscape. Journal of Wildlife Management. 2015;**79**:749-765. DOI:10.1002/jwmg.896

[26] Kendall KC. Grizzly and Black Bear Feeding Ecology in Glacier National Park, Montana: Progress Report. West Glacier (MT): Science Center, Glacier National Park; 1986

[27] McCall BS, Mitchell MS, Schwartz MK, Hayden J, Cushman SA, Zager P, et al. Combined use of mark-recapture and genetic analyses reveals response of a black bear population to changes in food productivity. Journal of Wildlife Management. 2013;**77**(8):1572-1582. DOI:10.1002/jwmg.617

[28] Prevéy JS, Parker LE, Harrington CA, Lamb CT, Proctor MF. Climate change shifts in habitat suitability and phenology of huckleberry (*Vaccinium membranaceum*). Agriculture and Forest Meteorology. 2020;**280**:1-12. DOI:10.1016/j.agrformat.2019.107803

[29] Pyne SJ. Fire: A Brief History. Seattle (WA): University of Washington Press; 2001. p. 204. https://ebookcentral-proquest-com.weblib.lib.umt.edu:2443/lib/msoumt/detail.action?docID=3444376

[30] Abatzoglu JT, Rupp DE, Mote PW. Seasonal climate variability and change in the Pacific Northwest of the United States. Journal of Climate. 2013;**27**:2125-2142. DOI:10.1175/JCLI-D-13-002181.1

[31] Dobkins R, Hummel SS, Lewis C, Pochis G, Dickey E. Tribes of the Oregon Country: Cultural plant harvests and indigenous relationships with ancestral lands in the twenty-first century. Oregon Historical Quarterly. 2017;**118**(4):488-517. DOI:10.5403/oregonhistq.118.4.0488

[32] Shores CR, Mikle N, Graves TA. Mapping a keystone shrub species, huckleberry (*Vaccinium membranaceum*), using seasonal colour change in the Rocky Mountains. International Journal of Remote Sensing. 2019;**40**(15):5695-5715. DOI:10.1080/01431161.2019.1580819

[33] Roberts DR, Nielsen SE, Stenhouse GB. Idiosyncratic responses of grizzly bear habitat to climate change based on projected food resource changes. Ecological Applications. 2014;**24**:1144-1154. DOI:10.1890/13-0829.1

[34] Lamb CT, Mowat G, McLellan BN, Nielson SE, Boutin S. Forbidden fruit: Human settlement and abundant fruit create an ecological trap for an apex omnivore. Journal of Animal Ecology. 2017;**86**:55-65. DOI:10.1111/1365-2656.12589

[35] Inouye DW. The ecological and evolutionary significance of frost in the context of climate change. Ecology Letters. 2000;**3**:457-463. DOI:10.1046/j.1461-0248.2000.00165.x

[36] Inouye DW. Effects of climate change on phenology, frost damage, and floral abundance of montane wildflowers. Ecology. 2008;**89**:353-362. DOI:10.1890/06-2128.1

[37] Augspurger CK. Reconstructing patterns of temperature, phenology, and frost damage over 124 years: Spring damage risk is increasing. Ecology. 2013;**94**:41-50. DOI:10.1890/12-0200.1

[38] Straka JR, Starzomski BM. Fruitful factors: What limits seed production of flowering plants in the alpine? Oecologia. 2015;**178**:249-260. DOI:10.1007/s00442-014-3169-2

[39] Boggs CL, Inouye DW. A single climate driver has direct and indirect effects on insect population dynamics. Ecology Letters. 2012;**15**:502-508. DOI:10.1111/j461-0248.2012.01766.x

[40] Bale JS, Hayward SAL. Insect overwintering in a changing climate. Journal of Experimental Biology. 2010;**213**:980-994. DOI:10.1242/jeb.037911

[41] Mainka SA, Howard GW. Climate change and invasive species: Double jeopardy. Integrated Zoology. 2010;**5**:102-111. DOI:10.1111/j.1749-4877.2010.00193.x

[42] Schuh RT, Slater JA. True Bugs of the World (Hemiptera:Heteroptera). Classification and Natural History. Ithica (NY): Comstock Publishing Associates; 1995. p. 336

[43] Langille AB, Arteca EM, Newman JA. The impacts of climate change on the abundance and distribution of the Spotted Wing Drosophila (*Drosophila suzukii*) in the United States and Canada. PeerJ. 2017, 2017;**5**:e3192. DOI:10.7717/peerj.3192

[44] McCloskey SPJ, Daniels LD, McLean JA. Potential impacts of climate change on western hemlock looper outbreaks. Northwest Science. 2009;**83**(3):225-238. DOI:10.3955/046.083.0306

[45] Donnelly A, Caffarra A, O'Neill BF. A review of climate-driven mismatches between interdependent phenophases in terrestrial and aquatic ecosystems. International Journal of Biometeorology. 2011;**55**:805-817. DOI:10.1007/s00484-011-0426-5

[46] Kudo G, Ida TY. Early onset of spring increases the phenological mismatch between plants and pollinators. Ecology. 2013;**94**(10):2311-2320. DOI:10.1890/12-2003.1

[47] Post E, Forchhammer MC. Climate change reduces reproductive success of an Arctic herbivore through trophic mismatch. Philosophical Transactions of the Royal Society of British Biological Sciences. 2008;**363**:2369-2375. DOI:10.1098/rstb.2007.2207

[48] Doiron M, Gauthier G, Levesque E. Trophic mismatch and its effects on the growth of young in an Arctic herbivore. Global Change Biology. 2015;**21**:4364-4376. DOI:10.1111/gcb.13057

[49] Kreyling J. Winter climate change: A critical factor for temperate vegetation performance. Ecology. 2010;**91**:1939-1948. DOI:10.1890/09-1160.1

[50] Wipf S, Rixen C. A review of snow manipulation experiments in the Arctic and alpine tundra ecosystems. Polar Research. 2010;**29**:95-109. DOI:10.1111/j.1751-8369.2010.00153.x

[51] Cooper EJ. Warmer shorter winters disrupt arctic terrestrial ecosystems. Ecological and Evolutionary Systems. 2014;**45**:271-295. DOI:10.1146/annurev-ecolsys-120213-091620

[52] Kudo G, Cooper EJ. When spring ephemerals fail to meet pollinators: Mechanisms of phenological mismatch and its impact on plant reproduction. Proceedings of the Royal Society B: Biological Sciences. 2019;**286**:20190573. DOI:10.6084

[53] Ogilvie JE, Griffin SR, Gezon KJ, Inouye BD, Underwood N, Inouye DW, et al. Interannual bumble bee abundance is driven by indirect climate impacts on floral resource phenology. Ecology Letters. 2017;**20**:1507-1515. DOI:10.1111/ele.12854

[54] Biesmeijer JC, Roberts PM, Reemer M, Ohlemuller R, Edwards M, Peeters T, et al. Parallel declines in pollinators and insect-pollinated plants in Britain and the Netherlands. Science. 2006;**313**:351-353. DOI:10.1126/science.1129551

[55] Isaacs R, Kirk AK. Pollination services provided to small and large highbush blueberry fields by wild and managed bees. British Ecological Society. 2010;**47**(4):841-849. DOI:10.1111/j.1365-2664.2010.01823.x

[56] Kearns CA, Inouye DW, Waser NM. Endangered mutualisms: Conservation of plant-pollinator interactions. Annual Review of Ecology and Systematics. 1998;**29**:83-112. DOI:10.1146/annurev.ecolsys.29.1.83

[57] Withgott J. Pollination migrates to top of conservation agenda. Bioscience. 1990;**49**(11):857-862. DOI:10.2307/1313643

[58] Thomann M, Imbert E, Devaux C, Cheptou PO. Flowering plants under global pollinator declines. Trends in Plant Science. 2013;**18**(7):353-359. DOI:10.1016/j.tplants.2013.04.002

[59] Graves TA, Janousek WM, Gaulke SM, Nicholas AC, Keinath DA, Bell CM, et al. Western bumble bee: Declines in the continental United States and range-wide information gaps. Ecosphere. 2020;**11**(6):1-13. DOI:10.1002/ecs2.3141

[60] Klein AM, Vaissier BE, Cane JH, Steffan-Dewenter I, Cunningham SA, Kremen C, et al. Importance of pollinators in changing landscapes for world crops. Proceedings of the Royal Society Bulletin. 2006;**270**(1608):955-961. DOI:10.1098./rspb.2006.3721

[61] Burd M. Bateman's principle and plant reproduction: The role of pollen limitation in fruit and seed set. Botanical Review. 1994;**60**(1):83-139. DOI:10.1007/bf02856594

[62] Aras P, DeOliveira D, Savoie L. Effect of a honey bee (Hymenoptera: Apidae) gradient on the pollination and yield of lowbush blueberry. Journal of Economic Entomology. 1996;**89**(5):1080-1083. DOI:10.1093/jee/89.5.1080

[63] Javorek S, Mackenzie K, Vander KS. Comparative pollination effectiveness among bees (Hymenoptera: Apoidea) on lowbush blueberry (Ericaceae: *Vaccinium augustifolium*). Annals of the Entomological Society of America. 2002;**95**:345-351. DOI:10.1603/0013-8746(2002)095%5B0345:CPEABH%5D2.0.CO;2

[64] Ratti CM, Higo HA, Griswold TL, Winston ML. Bumble bees influence berry size in commercial Vaccinium spp. cultivation in British Columbia. Canadian Entomologist. 2008;**140**(3):348-363. DOI:10.4039/n07-065

[65] MacKenzie KE. Pollination requirements of three highbush blueberry (Vaccinium corymbosum L.) cultivars. Journal of the American Horticultural Society. 1997;**122**(6):891-896. DOI:10.21273/jashs.122.6.891

[66] Rogers SR, Tarpy DR, Burrack HJ. Bee species diversity enhances productivity and stability in a perennial crop. PLoS One. 2014;**9**(5):e97307. DOI:10.1371/journal.pone.0097307

[67] Whitlock C, Cross W, Maxwell B, Silverman N, Wade AA. Montana Climate Assessment. Bozeman and Missoula, MT: Montana State University and University of Montana, Montana Institute on Ecosystems; 2017. p. 318. DOI:10.15788/m2ww8w

[68] USDA NRCS National Water and Climate Center. Report Generator 2.0. [Internet]. 2016. Available from https://wcc.sc.egov.usda.gov/reportGenerator/ [Accessed January 12, 2023]

[69] Akre RD, Greene A, MacDonald F, Landolt PJ, Davis HG. Yellowjackets of America North of Mexico. Washington, DC: U.S. Department of Agriculture. p. 102

[70] Irwin RE, Bronstein JL, Manson JS, Richardson L. Nectar robbing: Ecological and evolutionary perspectives. Annual Review of Ecology, Evolution and Systematics. 2010;**41**:271-292. DOI:10.1146/annurev.ecolsys.110308.120330

[71] Chacoff NP, Aschero V. Frequency of visits by ants and their effectiveness as pollinators of Condalia microphylla. Journal of Arid Environments. 2014;**2014**:91-91. DOI:10.1016/j.jaridenv.2014.03.001

[72] Larson B, Kevan P, Inouye D. Flies and flowers: Taxonomic diversity of anthophiles and pollinators. The Canadian Entomologist. 2001;**133**:439-465. DOI:10.4039/ent133439-4

[73] Cardinal S, Buchmann SL, Russell AL. The evolution of floral sonication, a pollen foraging behavior used by bees (Anthophila). Evolution. 2018;**72**(3):590-600. DOI:10.1111/evo.13446

[74] Chacoff NP, Aizen MA, Aschero V. Proximity to forest edge

does not affect crop production despite pollen limitation. Proceedings: Biological Sciences. 2008;**275**(1637):907-913. DOI:10.1098/rspb.2007.1547

[75] Kearns CA, Inouye DW. Techniques for Pollination Biologists. Niwot (CO): University of Colorado Press; 1993. p. 583

[76] Berman E, Graves TA, Mikle N, Merkle J, Johnston A, Chong G. Comparing quality and trend of remotely sensed phenology and productivity metrics across the Western United States. Remote Sensing. 2020;**12**(16):2538. DOI:10.3390/rs12162538

[77] Salish C, Tribes K. Climate Change Strategic Plan. Pablo (MT): Confederated Salish and Kootenai Tribes of the Flathead Nation; 2016. p. 86

[78] Thomson JD. Flowering phenology, fruiting success and progressive deterioration of pollination in an early-flowering genophyte. Proceedings of the Royal Society of London B. 2010;**365**:3187-3199. DOI:10.1098/rstb.2010.0115

[79] USFWS (United States Fish and Wildlife Service). Federal Register Documents Proposed Rules. 2016;**81**(51)

[80] Mola J, Hemberger J, Kochanski J, Richardson L, Pearse IS. The importance of forests in bumble bee ecology and conservation. Bioscience. 2008;**71**(12):1234-1248. DOI:10.1093/biosci/biab121

[81] Colla SR, Gadallah F, Richardson L, Wagner D, Gall L. Assessing declines of north American bumble Bees (Bombus spp.) using museum specimens. Biodiversity and Conservation. 2012;**21**:3583-3595. DOI:10.1007/s10531-0112-0383-2

[82] Goulson D, Hanley ME, Darvill B, Ellis JS, Knight ME. Causes of rarity in bumblebees. Biological Conservation. 2005;**122**:1-8. DOI:10.1016/j.biocon.2004.06.017

[83] Goulson D, Nicholls E, Botias C, Rotheray EL. Bee declines driven by combined stress from parasites, pesticides, and lack of flowers. Science. 2015;**347**:1435-1444. DOI:10.1126/science.1255957

[84] Janousek WM, Douglas M, Cannings S, Clement M, Delphia C, Everett J, et al. Recent and future declines in a historically widespread pollinator linked to climate, land cover, and pesticides. Proceedings North American Academy of Sciences. 2023;**120**(5):1-9. DOI:10.1073/pnas.2211223120

[85] Sirois-Delisle C, Kerr JT. Climate change-driven range losses among bumblebee species poised to accelerate. Scientific Reports. 2018;**8**:14464. DOI:10.1038/s41598-018.32665-y

Section 2

Disease and Control Methods in Strawberry, Raspberry, and Blueberry Production

Chapter 3

Raspberry and Blueberry Viruses

Melike Yurtmen

Abstract

Berry fruits are nutrient-rich foods and proved to be beneficial for human health. High nutraceutical properties of these fruits, in particular, blueberries and raspberries, such with exceptionally high antioxidant levels, fiber, and a low natural sugar content have an important criterion for the marketable value as well. Thus, they are defined as functional foods. Naturally, rapid increases in their production and consumption rates during recent years, therefore, make sense. In the cultivation of these berry crops, satisfactory yields and quality of the produce posed a threat by certain pests in the world. Blueberries and raspberries are also infected with viruses that have been identified as new viruses on top of existing ones as a result of expanding cultivation areas around the world. Incurable plant pathogenic viruses cause major damage to the members of blueberry and raspberry as if serious yield losses and longevity of plantings. This chapter is intended to compile knowledge of pathogenic plant viruses that infect blueberry and raspberry plants. Herein, a review of geographic distribution, importance, symptoms, transmission, detection, and management of these berry viruses are provided for the readers.

Keywords: pathogenic plant viruses, *Vaccinium*, *Rubus*, blueberry, raspberry, management strategies

1. Introduction

Blueberry and raspberry production is increased worldwide toward consumer demand over the last decades. Accordingly, cultivation of these fruit crops expanded to new areas in the world. Hereby, viruses infecting plants which prevent their growth and production increased in number significantly year by year since the beginning of this millennium. Pathogenic plant viruses pose a threat to their producers since no curative treatment exists [1]. However, not all viruses cause serious damage to plants, but some are widespread and destructive. Major losses in terms of quantity and quality of the crop are caused by these destructive viruses. Besides, members of the blueberry and raspberry plants are known to be hosting a wide range of pathogenic viruses which, especially when present as mixed infections, most probably cause severe symptoms and greatly affect the yield and viability of the plant. These viruses impact their growers by inducing weak stand establishment and fruit quality, reduced yield, and plant decline. However, symptoms caused by viral diseases in plants vary from being completely asymptomatic to dying of the plant. This symptomatic range is affected by several factors for instance weather conditions, the growing methods, and the age and type of the cultivars. At the beginning of all, to understand what the virus is, how it spreads, and how it is determined and controlled is utmost important to manage diseases induced by viruses. Considering all these aspects, knowledge of pathogenic plant viruses infecting blueberry and raspberry plants in particular is reviewed in terms of their geographic distribution, importance, symptoms, transmission, detection, and management strategies in this chapter.

2. Understanding plant viruses

Plant pathogenic viruses are highly infectious, submicroscopic, and obligate intercellular parasites with a DNA or RNA core. The nucleic acid is the infectious part of the virus surrounded by a protein coat to be able to replicate within the living cells, and the protein coat is the protective part of the particle. They are too small to be seen with the human eye, but they can be observed with electron microscopes. Their taxonomic classification is based on chemical compounds as well as morphologic, biologic, pathologic, and molecular features. The given names, which are in English, are taken from the plant that they infect, the unique symptoms they cause, and the agent that causes them.

Plant pathogenic viruses are immobile agents and need to be spread from infected to healthy plants or plantations via several means of transmission. Their transmission can be divided into two types and are called as vertical transmission and horizontal transmission.

Vertical transmission occurs when the progeny of an infected parent plant inherits the virus through seeds or pollen, as well as vegetative propagation. Horizontal transmission occurs between generations via fungi, invertebrate vectors (nematodes, insects, and mites), human pruning shears and tools, and other types of direct, external contamination [2, 3].

Aphids, whiteflies, thrips, and leafhoppers are the most important plant virus vectors because they have piercing–sucking mouth parts that allow the insects to reach and feed on the contents of plant cells. Among these, aphids and whiteflies are the most capable of transmitting virus species.

The virus is spread by sap-sucking insects in two ways: persistent transmission and non-persistent transmission. The time an insect takes to acquire and transmit a virus determines the differences.

Pathogenic plant viruses are classified into two types based on their spread distance from their origin. The transfer of contaminated propagation material allows viral infections to spread over long distances. The viral diseases can spread locally (short-distance movement) by pollen from infected plants or by direct contact by vectors such as nematodes, aphids, thrips, soil-borne fungi, and leafhoppers.

When a pathogenic virus enters a region or country, its host range determines its spread. The majority of plant viruses, on the other hand, have a diverse spectrum of alternate hosts.

Initially, to spot virus-induced host plant problems, it is necessary to understand what a healthy plant looks like! Some plants have characteristics or habits that, at certain phases of development, can be misinterpreted as disease symptoms. Plants can also develop virus-like symptoms in reaction to adverse weather conditions, soil mineral/nutrient imbalances, infection by non-viral pathogens, insect/mite/nematode pest damage, air pollution, pesticides, and other factors.

A virus's pathogenic potential is revealed by its capacity to infect one or more plant species and create observable symptoms. Plant symptoms are frequently used to define a viral-etiology disease and to locate infected plants in order to control the disease. When symptoms are specific to a disorder, visual inspection is usually straightforward. Many factors, such as virus strain, host plant cultivar/variety, infection period, and habitat, may influence the symptoms shown. Some viruses may provide no obvious signs or asymptomatic disease. Furthermore, different viruses can generate the same symptoms in the same host, or different strains of the same virus can cause different symptoms in the same host.

Plant viruses create significant economic losses and threaten sustainable agriculture. There are no antiviral chemicals available to heal plants once they have been infected by a viral pathogen. Control strategies that are effective can considerably reduce or prevent the disease from arising. In this step, virus identification is a required initial step in the management of a virus-caused disease. Plant viruses can be difficult to detect since the typical symptoms are similar to herbicide injury, air pollution damage, mineral deficiencies, and other plant diseases, or more than one virus infection. As a result, identifying viral infection is frequently required.

A critical component of every crop management system is the early and accurate detection of plant diseases. When it is determined that a virus is the cause of a disease, a set of tests are required to determine its identity. Even though identifying a virus or virus complex involves one or more other diagnostic techniques, symptom assessments play a significant part in the diagnostic process.

Understanding the biology of the virus such as virus properties, host range, vector, and inoculum sources is required for effective disease management. It is difficult to eliminate pathogens without harming the host plant due to their nature. As a result, the majority of management techniques for plant viruses are aimed at preventing plant infection.

This page discusses blueberry and raspberry viral infections in light of this understanding.

3. Viruses infecting blueberry (*Vaccinium corymposum* L.; *V. augustifolium* Ait. and *V. ashei* Reade)

3.1 Blueberry fruit drop associated virus (BFDaV)

The virus is a member of the genus Vaccinivirus in the family *Caulimoviridae*. BFDaV has a DNA genome and can detect through polymerase chain reaction (PCR). The only host of BFDaV is blueberry (*Vaccinium corymbosum*). As a disease, fruit dropping of blueberry was first detected in Canada (British Colombia) in the late 1990s. The virus was then observed in the USA in 2012. Young leaves show the red coloration of the veins, and the corolla of the flowers exhibits red striping while blooming. These symptoms stay only in the blooming period, and then, the plants appear as if normal. More symptoms are described as aborting the fruits (3–5 mm in diameter) around 100% in infected bushes nearly 3 weeks before harvest. Therefore, during the harvest, infected blueberry plants look taller compared to healthy ones [4].

3.2 Blueberry latent spherical virus (BLSV)

In Japan, Blueberry latent spherical virus (BLSV) was isolated from an asymptomatic highbush blueberry. It is a Nepovirus [5]. It can detect by reverse transcription-PCR (RT-PCR). The virus was reported as present in blueberries but with no symptom expression. Thus, the virus infections call as latent in blueberry trees. BLSV epidemiology and mode of transmission have not been determined yet.

3.3 Blueberry latent virus (BBLV)

Blueberry latent virus is present in the USA, Canada, and Japan. It is symptomless and does not have negative effects in mixed infections. As a result, the virus's prevalence is of minor concern. It is transmissible by pollen and seed. Taxonomically, it belongs to the genus Amalgavirus and is possible to detect by RT-PCR [6].

3.4 Blueberry leaf mottle virus (BLMoV)

Blueberry leaf mottle virus is reported from blueberry and its hybrids from Canada and USA. It spreads by seeds and pollen—1.5% of blueberry seedlings from a contaminated bush [7]. Although it is a Nepovirus, no means of nematode transmission has yet been identified. It occurs naturally in highbush blueberry (*V. corymbosum*) and its hybrids with a grapevine (*Vitis vinifera, Vitis labrusca*).

The highbush blueberry cultivars 'Jersey' and 'Rubel' are affected negatively by this virus disease. The onset of symptoms is monitored after 4 years of the latent period from the initial infection. Stem dieback and stunting are developing symptoms of infected bushes. Malformed and mottled leaves caused by the shortening internodes appear as if piled on top of one another. Pale yellow-green leaves are smaller than the ones on uninfected bushes. Nearly no yield is harvested [8]. Virus presence was reported in wild species of blueberries. Detection of the virus by ELISA and/or RT-PCR is available.

3.5 Blueberry red ringspot virus (BRRSV)

Red ringspot disease has been described by Hutchinson [9], and Hutchinson & Varney [10]. It is then assigned to the Soymovirus genus with a DNA genome in the family *Caulimovididae*. Highbush blueberry (*V. corymbosum* L.) is the host, and symptom expressions are observed on leaves, stems, and fruits [11]. The virus is present in the USA, Japan, the Czech Republic, Slovenia, and Poland [12–16]. It is a dsDNA virus and is available to diagnose by PCR [17, 18]. It is found in nature only in *Vaccinium* spp. The only known mode of transmission is vegetative propagation.

Very characteristic red ring spots are observed on at least one older stem. Mature leaves have 2 to 6 mm diameter pale green-centered reddish-brown circular markings. In mid to late summer, these symptoms appear on the upper surface of the leaf. This is a diagnostic symptom, but sometimes rings are visualized on both sides of the leaf in some cultivars. Red-stemmed or dark cultivars may have masked on matured stems [19].

Reddish rings appear on green fruits and hide as they ripen. Small, deformed, and late-ripening berry fruits are distinguishing symptoms of BRRSV.

3.6 Blueberry shoestring virus (BlSSV)

Blueberry shoestring virus (BSSV) belongs to the genus Sobemovirus, an RNA-containing virus, transmissible by aphids. The only known host of BlSSV is blueberries. It is present in the USA and Canada [20, 21]. The disease can be diagnosed by enzyme-linked immunosorbent assay (ELISA) and RT-PCR. After 4 years of latent period, symptoms are recorded. The spread of the virus in an infected blueberry area occurs horizontally from one bush to another [22]. Elongated (0.2–1.2 cm) reddish stripes that disappear by the growing season are the most distinct symptom on the stems. Flower breaks, strap-like, narrow, and curled leaves are also observed. A significant reduction in the yield of infected bushes has been reported [21–24]. Disease resistance is known for 'Blueray' and 'Atlantic' cultivars of blueberries [25]. BSSV is cited under the most common highbush blueberry (*V. corymbosum*) viruses. Yield losses may reach up to 25% of infected bushes [15]. In 1981, a $3 million yield loss caused by BlSSV has been recorded in the USA. One of the most significant economic damages to highbush blueberries is caused by BlSSV [14]. Mechanical inoculation of the virus on blueberry seedlings or rooted softwood cuttings is possible [20].

3.7 Blueberry virus A (BVA)

BVA was first reported in Japan in the 'Spartan' highbush blueberry cultivar. The presence of the virus is known in Canada and the USA. Blueberry virus A has a latent infection in blueberries [26, 27]. It is assigned to the family *Closteroviridae*, but not yet to a genus. Detection is available by RT-PCR [26]. BVA has no specific symptom expression in single infections. Therefore, no data is available on the economic damage to the blueberry industry.

3.8 Blueberry mosaic associated virus (BlMaV)

The mosaic disease of blueberries has been recorded in the 1950s and is suspected to be a virus since it was transmitted by grafting [23]. It was reported not only from mosaic-affected blueberries but also from symptomless plants.

Yellow, yellow to green, and pink coloring, mosaic and mottling of the leaves, late ripening with low quality of the fruits, and reduction in yield are the symptoms induced by BlMaV [23, 28]. BlMaV as a member of the family *Ophioviridae* was detected in tested plants with symptoms [29]. The presence of the virus is reported in North and South America, Asia, Europe, New Zealand, and South Africa [27, 30, 31]. Blueberry mosaic spreads slowly in the field by unknown means. This disease is of relatively little concern.

3.9 Blueberry necrotic ring blotch virus (BNRBV)

The virus (BNRBV) was reported first with necrotic ring blotch symptoms in highbush blueberries in the USA in 2006 [27]. It is a new genus, Blunervirus, with an RNA genome [32, 33]. Detection of the virus is available by RT-PCR. IBlueberries infected with the virus displayed distinct necrotic rings with green cores, but when the rings joined. Earlier defoliation can be observed on severely infected bushes that are confused with Septoria leaf spot disease. BNRBV, unlike BRRV, infects all the leaf surfaces without any stem symptoms.

3.10 Blueberry scorch virus (BlScV)

In the 1980s, it is found in blueberries in the USA. No symptoms occur on tolerant cultivars. However, necrosis on flowers and young leaves is observed on sensitive cultivars. Die-back of twigs has also been recorded. The degree of blighting severity determines the size of the bearing fruits. Twig die-back allows lateral buds to develop and form branches later in the season below the point of necrosis. The productivity of infected plants that show symptoms declines year by year, and finally, the plants die [34]. BlScV can cause a serious economic impact due to loss of yield and premature death of plants. Aphid vector spreads the virus in nature [34]. The virus belongs to the Carlavirus genus with an RNA genome. Diagnostic techniques based on ELISA and RT-PCR are also available for virus detection [35].

3.11 Blueberry green mosaic-associated virus (BGMaV)

The virus was discovered first in 2006 as a novel Vitivirus from a blueberry plan showing green mosaic symptoms [27]. It is named as 'blueberry green mosaic-associated virus' (BGMaV). It has an RNA genome, and its vector is unknown [36, 37]. Green-centered necrotic rings can also be confused with the ones induced by fungal pathogens in blueberries [27]. It is detectable by RT-PCR and presents only in the USA [27].

3.12 Blueberry shock virus (BlShV)

Shock symptoms of blueberries have been reported first in 1987 in the USA and then correlated with Blueberry Scorch Virus. It is an Ilarvirus and is transmitted by pollen. BlShV moves by wind and bees during pollination [38]. Diagnosis of the virus is available by ELISA or RT-PCR.

After the initial infection, symptom expressions take up to 2 years. Bushes demonstrate a 'shock reaction' the year after infection, with flowers and foliage blighting in early spring. Sudden dieback of young vegetative shoots and flowers, necrosis and blighting in flowers, defoliation, and lack of fruit set are the symptoms observed. Plants may not show symptoms in spring growth following the first symptom expression. Partially blighted bushes show symptoms the year after although they were symptomless wood the year before. Infected blueberries normally bring into flowers after 1–3 years and bear fruits as if normal although they are still infected and the source of inoculum [34, 38–40].

3.13 Peach rosette mosaic virus (PRMV)

Rosette mosaic disease of peaches was discovered as the virus in the 1970s. It is then reported from grapes and blueberries [41–43]. Symptoms observed in blueberries are distorted, malformed, and uneven distribution of bushes in the plant. Yield losses due to the PRMV infection in blueberries are not known. The presence of PRMV is recorded in Canada and USA. It is an RNA virus and is classified as a Nepovirus with soil-borne nature [44–46]. Virus detection is available by ELISA and RT-PCR.

4. Viruses infecting raspberry (*Rubus idaeus* L. and *Rubus occidentalis* L.)

4.1 Apple mosaic virus (ApMV)

ApMV is in the Ilarvirus genus with an RNA genome and has a wide variety of host plants in the *Rosaceae* family worldwide [47]. It spreads by pollen. The presence of ApMV is detected in *R. idaeus R. occidentalis* and *R. ursinusin* [48–50]. The virus is symptomless in *Rubus*. However, some *R. idaeus* plants in Germany showed symptoms of yellow mottling and/or line patterns [48]. Virus detection is available by ELISA and RT-PCR.

4.2 Arabis mosaic virus (ArMV)

ArMV was discovered for the first time in the 1940s [51]. The virus infects about 100 plant species from roughly 30 families, producing considerable losses in the majority of them [52, 53]. It is a Nepovirus that is spread by nematode vectors [54, 55]. ARMV can be detected using ELISA and RT-PCR. Raspberry cultivars have genetic resistance to ArMV [56].

4.3 Black raspberry necrosis virus (BRNV)

The virus was first detected in black raspberry (*R. occidentalis* L.) in 1955 [57] by inducing latent or mild symptoms in red raspberry (*R. idaeus* L.) cultivars and severe cane tip necrosis in black raspberry [58]. It belongs to the *Secoviridae* family with an unassigned genus. Its transmission in nature occurs by the raspberry aphids and is distributed worldwide where raspberries are produced [59]. RT-PCR is available for virus detection. *Rubus* is the genus hosting BRNV with its wild and cultivated species. Although BRNV is symptomless in several commercial species, it is still on the list of the most important viral pathogens of *Rubus* spp. Some BRNV-infected red raspberry cultivars show chlorotic spots and venial chlorotic mottle in leaves, but most of them are symptomless carriers of the virus. Symptoms caused by the virus can be listed as leaf chlorosis, mottling, and puckering. Infected plants do not yield fruits within 3–4 years [60]. Shorter and thinner canes and smaller fruits are also induced by the virus in sensitive cultivars. Yield loss caused by the virus decreases up to 30% in sensitive cultivars [61].

4.4 Cherry rasp leaf virus (CRLV)

CRLV is a type member of the Cheravirus genus [62]. It is spread by the nematode. RT-PCR can be used to diagnose the virus. The virus is found naturally in North America. Rubus has only been identified in a few red raspberries shipped from Canada to Scotland [63]. There is no information concerning the economic importance of the virus in commercial *Rubus* crops.

4.5 Cucumber mosaic virus (CMV)

CMV occurs worldwide in many different plant species [64], including *Rubus* spp. The virus is a member of the Cucumovirus genus with an RNA genome and is transmitted by many aphid species in nature [65]. CMV is also seed transmitted in several of its hosts [64], but not in raspberry seedlings. Detection of the virus is available by ELISA and RT-PCR. It is reported in *Rubus* only from Britain and Eastern Russia. The first report of the virus is from a few plants of *R. idaeus* L. cv. 'Lloyd George' in Scotland [66]. The second record of the virus comes from Scotland [67] and the Soviet Far East [68]. The virus is symptomless in cultivated brambles [67]. It has mild foliar symptoms in red raspberries although it is lethal in *R. phoenicolasius* Maxim [66–68]. Some raspberry cultivars show chlorotic mottling and blotching while red raspberry cv. 'Lloyd George' has pale green blotching of the leaves with no apparent effect on plant vigor or fruiting [66]. Foliar chlorotic ringspot symptoms are recorded by the single CMV infection of red raspberry with no obvious degeneration in vigor. Small leaves with bright chlorotic mottling were another symptom detected in infected raspberries (cv. 'Visluha') [68]. However, infected *R. phoenicolasius plants* show chlorotic blotch and line patterns in leaves which become bright yellow in summer, decline in vigor, and death of the plants in 3–4 years [67].

4.6 Raspberry bushy dwarf virus (RBDV)

RBDV belongs to the Idaeovirus genus with an RNA genome. It is a pollen and seed-borne virus [69]. Laboratory tests are available by ELISA and RT-PCR. The virus infects many *Rubus* species and cultivars worldwide. There are resistance-breaking (RB) isolates of the virus infecting raspberry cvs (Glen Clova, Malling Admiral, Malling Delight, Malling Jewel, Willamette, Haida) which are known to be immune to the common strain of RBDV [70, 71]. The RB isolates are recorded from Europe. Pollen-borne nature of the virus causes drupelet abortion thus crumbly fruit forms in some red raspberry cultivars. RBDV infection alone is symptomless in red raspberry cultivars. However, crumbly fruit disease can be observed in mixed infections. Mixed infection of RBDV with BRNV induces dwarfing and shoot proliferation in red raspberries, a typical bushy dwarf condition. The origin of virus name is derived from mixed infection of the plants with RBDV and BRNV [72].

4.7 Raspberry latent virus (RpLV)

The virus is not yet assigned taxonomically. It is an aphid-borne virüs with an RNA genome and is detectable by RT-PCR [56]. It was identified from a red raspberry cv. Glen Prosen with leaf spot symptoms and therefore was named before as raspberry leaf spot virus in 1988 [73]. A typical symptom induced by the virus is the formation of a few, stunted canes which are very prone to autumn fruiting.

4.8 Raspberry leaf blotch virus (RLBV)

RLBV as a member of the Emaravirus genus with an RNA genome is readily diagnosed by RT-PCR. It has been originally described in the red raspberry cv. Glen Ample in Scotland and Serbia [74]. It is responsible for leaf blotch disorder which was found before in Tayberry and correlated with the infestation of plants with the raspberry leaf and bud mite, *Phyllocoptes gracilis* [75]. This virus causes symptoms such as yellow and light green leaf blotches and patches, distortion of leaf margins, and leaf twisting. However, these symptoms were associated with the feeding damage of mites which is widespread in Europe and North America.

4.9 Raspberry leaf curl virus (RpLCV)

RpLCV is reported first in the 1920s [76] and has hosts limited by the genus *Rubus*. The virus so far is not yet characterized. It is an aphid-borne virus. No laboratory diagnostic tool is available for the virus. The presence of the virus is reported only in the United States and Canada. Infected red raspberries show severe symptoms of curled, distorted, and chlorotic leaves in the following growing season right after infection. Shoots with extreme shortening of the internodes are dwarfed. Crumbly, seedy or small fruits are set in infected plants and naturally, plant yield is severely reduced. Since the virus spread is very slow in the field, there is no significant loss in terms of economic value [56, 77].

4.10 Raspberry leaf mottle virus (RLMV)

The virus is an aphid-borne Closterovirus with an RNA genome and is readily detectable by RT-PCR. It was first identified in 1924 as producing 'raspberry mosaic disease' (RMD). RLMV is a widespread virus in Europe and North America. It has a latent infection in most of the red raspberry cultivars although some diagnostic symptoms are induced in a few of them. It is the member of a disease complex inducing mosaic disease in raspberries along with rubus yellow net virus (RYNV) and black raspberry necrosis virus (BRNV). Additionally, RLMV is responsible for producing "raspberry crumbly fruit" syndromes together with the raspberry latent virus (RpLV) and raspberry bushy dwarf virus (RBDV) in a disease complex [56].

4.11 Raspberry ringspot virus (RpRSV)

RpRSV is a Nepovirus with an RNA genome transmissible by pollen, seed, and nematode [78]. Laboratory tests are available by ELISA and RT-PCR. It was first described in the 1950s as the causal agent of the raspberry leaf curl disease [79]. The virus is present in Europe and has a very wide range of host plants belonging to at least 14 families [80]. The virus resistance has been reported in red raspberries [56].

4.12 Raspberry vein chlorosis virus (RVCV)

The virus as a member of the Rhabdovirus genus with an RNA genome has an aphid vector [81] and is detectable by RT-PCR. RVCV was first described in 1952 and is widespread in New Zealand, the UK, and Europe, causing stunted cane growth and reduced vigor thereby reducing yield in particular with mixed infection. A notable symptom caused by the virus is characteristic chlorosis of the minor leaf veins in field-grown red raspberry plants.

4.13 Rubus yellow net virus (RYNV)

RYNV is an aphid-transmitted Badnavirus with a DNA genome and is detectable by PCR [82]. RYNV is a member of a disease complex inducing mosaic disease in raspberries along with raspberry leaf mottle virus (RLMV) and black raspberry necrosis virus (BRNV). The virus infects all the cultivars of red raspberries and most of the cultivars and hybrids of blackberries in North America and Europe [61]. It is symptomless in all red raspberries and although veinal chlorotic mottle or line-pattern symptoms may be observed in some of them. Yield losses of 30–75% in the first year and up to 15% in following years have been documented as a result of the combined infection with the black raspberry necrosis virus [83]. The rapid spread of the virus is correlated with the vector aphid population [56].

4.14 Sowbane mosaic virus (SoMV)

It is a pollen and seed-borne virus with an RNA genome belonging to the genus Sobemovirus. Laboratory tests are available by ELISA and RT-PCR. It was named first as Rubus chlorotic mottle virus (RuCMV) but then classified as Rubus strain of sowbane mosaic virus (SoMV-R). It was found recently in Scotland, in red raspberries and wild blackberries (*Rubus fruticosus*) [84]. *Rubus* strain of the virus is not seed transmitted. It is widespread [85]. Identical symptoms produced by SoMV-R are the reverse curling of the tip leaves in *R. idaeus* 'Gaia' temporarily and diffuse chlorotic spots in blackberry leaves.

4.15 Tomato black ring virus (TBRV)

It is transmitted by pollen, seed, and nematode and belongs to the genus Nepovirus with an RNA genome. Laboratory diagnosis of the virus is available by ELISA and RT-PCR. It is described in 1946 [25, 86] and has a very large host range including *Rubus* spp. It is present in Europe, Japan, India, and Saudi Arabia. The virus together with RpRSV in mixed infection causes diseases named raspberry leaf curl or raspberry ring spot based on the cultivar. The virus's host range is as broad as that of other nematode-transmitted viruses. Resistance to TBRV has been identified in blackberry and raspberry cultivars [56].

5. Common viruses infecting blueberry and raspberry

5.1 Cherry leaf roll virus (CLRV)

The virus is a member of the Nepovirus genus with an RNA genome and is detectable by ELISA and RT_PCR [87]. It is a pollen and seed-borne virus and is transmissible by nematodes. It was found in blackberies first in England [88]. It was reported as sometimes being lethal in *Rubus armeniacus* 'Himalaya Giant'. The second record came from cultivated red raspberries in New Zealand showing vigor depression and severe leaf symptoms [89]. The virus is reported with a wide natural host range around the world [90]. Chlorotic mottling and line patterning in leaves are produced in the blackberry cv. Himalaya Giant by CLRV. CLRV-infected red raspberry cultivars have stunted and deformed leaves with severe chlorotic mottle and ring and line patterns.

5.2 Strawberry latent ringspot virus (SLRSV)

It is an RNA virus and not yet assigned species in the *Secoviridae* family [91]. The virus is detectable by ELISA and RT-PCR. SLRSV has a very wide host range including more than 125 plant species belonging to 27 families and is transmissible by nematodes, pollen, and seed [92]. Yellowing and stunting in blackberry and raspberry are the symptoms observed. The virus is reported from Rubus plants only in Europe. Blackberry and raspberry cultivars have the sources of resistance to SLRSV [56].

5.3 Tobacco ringspot virus (TRSV)

TRSV is a member of the genus Nepovirus with an RNA genome [93, 94]. Laboratory analyses are available by ELISA and RT-PCR. It is a pollen and seed-borne virus transmitted by nematodes. TRSV has a worldwide distribution. It was identified first in 1917 [95]. The presence of the virus in *Rubus* is reported in the USA, from wild blackberries [96]. TRSV in single infections produces mild or no symptoms. However, severe symptoms are produced when co-infection occurs with other viruses. The only resistant cultivar to the common strain of the virus is reported as 'Jersey'.

5.4 Tomato ringspot virus (ToRSV)

The virus is classified in the Nepovirus genus and has an RNA genome. It is transmitted by nematodes and is detectable by ELISA and by RT-PCR. The first isolation of the virus in *Rubus* was in 1938 in Canada [56]. The virus has a wide host range from 35 families [97]. In raspberry production, the virus can induce a serious problem. Blackberries can be infected by ToRSV as well [98, 99]. In raspberries, canes were partially or completely killed 3 years after becoming infected within 10–80% [100]. Infected raspberry plants have no symptoms in the first year. Meanwhile, following spring, some of the primocanes exhibit yellow rings, line patterns, or vein chlorosis in the leaves which may be called shock reactions although the symptoms are infrequent in the following years. The disease's red raspberry symptoms vary according on the cultivar, the length of the infection, and the stage of growth of the plants. Fruiting canes are delayed, with varied degrees of chlorosis on the leaves and a large proportion of deformed or crumbly fruit, as well as an overall fall in vigor as output decreases [101].

6. Management of virus diseases

There are a few alternatives for reducing the impact of plant virus diseases [102], but no anti-viral cure that block or interfere with virus infection have been identified. It is, therefore, plant virus disease management strategies based on conventional and non-conventional methods. In conventional methods, pest control, cultural control, development of virus-free/virus-tested plants, quarantine regulations, and breeding programs are employed.

Pathogen resistance is being employed in non-conventional ways. Coat protein, movement protein, replicase, satellite RNA, and antisense RNA all contribute in pathogen-derived resistance.

Insecticides are used to suppress viral vector populations.

Monitoring for virus-like symptoms, removing and burning of infected plants, and identifying alternate and reservoir hosts are all cultural techniques. Plants that show signs of infection should be inspected on a regular basis and removed. Remove weeds and plant debris that may harbor the disease. They should not be buried or composted.

The distribution of healthy plant material is a key step in preventing the spread of harmful viral infections that might accompany plant germplasm mobility. Alternative hosts act as virus reservoirs and cultural controls, as well as having a substantial impact on epidemiology.

Blueberry and raspberry plants are vegetatively multiplied perennial crops. One of the greatest ways for reducing the disease effect generated by pathogenic plant viruses is to establish orchards using virus-tested stock that is devoid of targeted viruses [1]. Apical meristem culture and/or chemotherapy/thermotherapy are used to generate virus-free/virus-tested category plants.

Quarantine regulations is a must to be followed during plant material import and export. Plant quarantine is a bio-security tool used to restrict the introduction and spread of economically relevant pests of plants or plant products that are not yet present in an area or are present but are not extensively distributed and are under official control [103, 104]. Consumer demand for these fruits has led to a rise in global output in recent years. Viral infections are easily spread since these crops are typically propagated and disseminated as vegetative cuttings.

Traditional breeding is used to create virus-tolerant/resistant cultivars; however, it is a time-consuming process that is not always practical or available.

The key ingredients in deciding which control measures to use and whether to use them alone or in combination with others are thorough epidemiological knowledge of the pathosystem in question, as well as solid information on the selectivity, mode of action, effectiveness, and reliability of each measure, as well as how to respond. To be adopted, control systems must be ecologically and socially sustainable, resilient, low-cost, and consistent with ordinary farming operations.

Virus epidemics in cultivated plants provide a global challenge to achieving acceptable yields and product quality. To confront this challenge, a more sophisticated and broad set of host resistance, cultural (phytosanitary and agronomic), chemical, biological, and legislative control techniques are becoming accessible. Knowing which elements limit viral epidemics in natural plant communities and primitive subsistence farming systems has aided in their evolution.

In Brief:

1. It is strongly advised to use certified, micro-propagated, virus-tested, and if possible, disease-free stock to establish nursery blocks and commercial plantings.

2. To control nematode-transmissible viruses, pre-plant soil testing for the presence of nematodes is advised, followed by pre-plant fumigation with an approved nematicide if vector nematodes are identified.

3. To eliminate new, latent infections, remove diseased plants as well as symptomless plants beyond the symptomatic plants in each direction.

4. Chemicals used to manage the vector, such as those targeting the mite or insect, may aid in reducing losses due by virus infection.

5. Weed removal to reduce possible in-season and overwintering viral reservoirs.

6. When available, choose resistant or tolerant cultivars or get transplants from a trustworthy supplier.

7. Disinfect all tools and machinery before and after use (one part bleach to four parts water).

Tobacco should not be used near vulnerable plants, especially in greenhouses. Cigarettes and other tobacco products may be contaminated and spread disease.

DOI: http://dx.doi.org/10.5772/intechopen.1001860

Author details

Melike Yurtmen
Biological Control Research Institute, Adana, Türkiye

*Address all correspondence to: melike.yurtmen@tarimorman.gov.tr

References

[1] Gergerich RC, Welliver RA, Gettys S, Osterbauer NK, Kamenidou S, Martin RR, et al. Safeguarding fruit crops in the age of agricultural globalization. Plant Disease. 2015;**99**:176-187

[2] Singh S, Awasthi LP, Jangre A. Transmission of plant viruses in fields through various vectors. In: Awasthi LP, editor. Applied Plant Virology. Cambridge, MA, USA: Academic Press; 2020. pp. 313-334

[3] Lefeuvre P, Martin DP, Elena SF, Shepherd DN, Roumagnac P, Varsani A. Evolution and ecology of plant viruses. Nature Reviews. Microbiology. 2019;**17**:632-644

[4] Diaz-Lara A, Martin RR. Blueberry fruit drop-associated virus: A new member of the family Caulimoviridae isolated from blueberry exhibiting fruit drop symptoms. Plant Disease. 2016;**100**:2211-2214

[5] Isogai M, Tatuto N, Ujiie C, Watanabe M, Yoshikawa N. Identification and characterization of blueberry latent spherical virus, a new member of subgroup C in the genus Nepovirus. Archives of Virology. 2012;**157**:297-303

[6] Saad N, Olmstead JW, Jones JB, Varsani A, Harmon PF. Known and new emerging viruses infecting blueberry. Plants. 2021;**10**:2172. DOI: 10.3390/plants10102172

[7] Childress AM, Ramsdell DC. Detection of blueberry leaf mottle virus in highbush blueberry pollen and seed. Phytopathology. 1986;**76**:1333-1337

[8] Ramsdell DC, Stace-Smith R. Blueberry leaf mottle, a new disease of highbush blueberry. Acta Horticulturae No. 95. In: II International Symposium on Small Fruit Virus Diseases, Budapest (Hungary). 3 July 1979. pp. 37-48

[9] Hutchinson MT. Can you recognize the symptoms of stunt disease? Proceedings of the Nineteenth Annual Blueberry Open House. 1950;**19**:9-11

[10] Hutchinson MT, Vamey EH. Ringspot: A virus disease of cultivated blueberry. Plant Disease Reporter. 1954;**38**:260-263

[11] Gillett JM. Physical and chemical properties of blueberry red ringspot virus. [M.S. thesis], Michigan State University, East Lansing, MI, USA. 1988

[12] Isogai K, Ishii M, Umemoto S, Watanabe M, Yoshikawa N. First report of blueberry red ringspot disease caused by blueberry red ringspot virus in Japan. Journal of General Plant Pathology. 2009;**75**:140-143

[13] Cho IS, Chung BN, Cho JD, Choi GS. First report of blueberry red ringspot virus infecting highbush blueberry in Korea. Plant Disease. 2012;**96**:1074

[14] Kalinowska E, Paduch-Cichal E, Chodorska M, Sala-Rejczak K. First report of blueberry red ringspot virus in highbush blueberry in Poland. Journal of Plant Pathology. 2011;**93**:S4.73

[15] Pleško IM, Marn MV, Koron D. Detection of blueberry red ringspot virus in highbush blueberry cv. 'Coville' in Slovenia. Julius-Kuhn-Archives. 2010;**427**:204-205

[16] Přibylová J, Špak J, Kubelková D, Petrzik K. First report of blueberry red ringspot virus in highbush blueberry in the Czech Republic. Plant Disease. 2010;**94**:1071

[17] Glasheen BM, Polashock JJ, Lawrence DM, Gillett JM, Ramsdell DC, Vorsa N, et al. Cloning, sequencing, and promoter identification of blueberry red ringspot virus, a member of the family Caulimoviridae with similarities to the "soybean chlorotic mottle-like" genus. Archives of Virology. 2002;**147**:2169-2186

[18] Polashock JJ, Ehlenfeldt MK. Molecular detection and discrimination of blueberry red ringspot virus strains causing disease in cultivated blueberry and cranberry. Plant Disease. 2009;**93**:727-733

[19] Cline WO, Ballington JR, Polashock JJ. Blueberry red ringspot observations and findings in North Carolina. Acta Horticulturae. 2009;**810**:305-312

[20] Ramsdell DC. Physical and chemical properties of blueberry shoestring virus. Phytopathology. 1979;**69**:1087-1091

[21] Lesney M, Ramsdell D. Purification and some properties of blueberry shoestring virus. Acta Horticulturae. 1976;**66**:105-109

[22] Caruso FL, Ramsdell DC. Compendium of Blueberry and Cranberry Diseases; American Phytopathological Society: St. MN, USA: Paul; 1995

[23] Varney EH. Mosaic and shoestring virus diseases of cultivated blueberry in New Jersey. Phytopathology. 1957;**47**:307-309

[24] Martin RR, Bristow PR. Scorch. In: Caruso FL, Ramsdell DC, editors. Compendium of Blueberry and Cranberry Diseases. St. Paul, MN, USA: American Phytopathological Society; 1995. pp. 51-52

[25] Murant AF. Tomato black ring virus. In: CMI/AAB Descriptions of Plant Viruses No. 38. Wellesbourne, UK: Association of Applied Biologists; 1970

[26] Isogai M, Muramatu S, Watanabe M, Yoshikawa N. Complete nucleotide sequence and latency of a novel blueberry-infecting closterovirus. Journal of General Plant Pathology. 2013;**79**:123-127

[27] Martin RR, Polashock JJ, Tzanetakis IE. New and emerging viruses of blueberry and cranberry. Viruses. 2012;**4**:2831-2852

[28] Ramsdell D, Stretch AW. Blueberry mosaic. In: Converse RH, editor. Virus Diseases of Small Fruits. Washington, DC, USA: US Department of Agriculture, Agriculture Handbook No. 631; US Government Printing Office; 1987. pp. 119-120

[29] Thekke-Veetil T, Ho T, Keller KE, Martin RR, Tzanetakis IE. A new ophiovirus is associated with blueberry mosaic disease. Virus Research. 2014;**189**:92-96

[30] Martin R, Tzanetakis I, Caruso F, Polashock J. Emerging and reemerging virus diseases of blueberry and cranberry. Acta Horticulturae. 2009;**810**:299-304

[31] Isogai M, Matsuhashi Y, Suzuki K, Yashima S, Watanabe M, Yoshikawa N. Occurrence of blueberry mosaic associated virus in highbush blueberry trees with blueberry mosaic disease in Japan. Journal of General Plant Pathology. 2016;**82**:177-179

[32] Quito-Avila DF, Brannen PM, Cline WO, Harmon PF, Martin RR. Genetic characterization of blueberry necrotic ring blotch virus, a novel RNA virus with unique genetic features. The Journal of General Virology. 2013;**94**:1426-1434

[33] Melzer MJ, Freitas-Astúa J, Rodriguez JCV, Roy A, Wei G. Create one new family Kitaviridae comprising three previously unassigned genera, Cilevirus, Blunervirus and Higrevirus. 2018;**002**. Available from: https://talk.ictvonline.org/ictv/proposals/2018.002P.A.Kitaviridae.zip. [Accessed: 04 July, 2023]

[34] Bristow PR, Martin RR, Windom GE. Transmission, field spread, cultivar response, and impact on yield in highbush blueberry infected with *blueberry scorch virus*. Phytopathology. 2000;**90**:474-479

[35] Wegener LA, Punja Z, Martin R, Bernardy M, MacDonald L. Epidemiology and identification of strains of blueberry scorch virus on highbush blueberry in British Columbia, Canada. Canadian Journal of Plant Pathology. 2006;**28**:250-262

[36] Thekke-Veetil T, Ho T. Molecular characterization of a new vitivirus discovered in a blueberry plant with green mosaic symptoms. Archives of Virology. 2019;**164**:2609-2611

[37] Thekke-Veetil T, Ho T, Polashock J, Tzanetakis I. Blueberry green mosaic symptoms are associated with the presence of a new vitivirus. In: Phytopathological. Vol. 106, No. 12. USA: Amer Phytopathological Soc.; 30 December 2016. p. 126

[38] Bristow PR, Windom GE, Martin RR. Recovery of plants infected with blueberry shock ilarvirus (BlShV). Acta Horticulturae. 2002;**574**:85-89. DOI: 10.17660/ActaHortic.2002.574.10

[39] MacDonald SG, Martin RR, Bristow PR. Characterization of an ilarvirus associated with a necrotic shock reaction in blueberry. Phytopathology. 1991;**81**:210-214

[40] Schilder A. Blueberry Shock Virus, Pest Alert Factsheet. East Lansing, MI, USA: Michigan State University Extension; 2009

[41] Ramsdell D, Gillett J. Peach rosette mosaic virus in highbush blueberry. Plant Disease. 1981;**65**:757-758

[42] Dias HF, Cation D. The characterization of a virus responsible for peach rosette mosaic and grape decline in Michigan. Canadian Journal of Botany. 1976;**54**:1228-1239

[43] Ramsdell D, Myers R. Peach rosette mosaic virus, symptomatology, and nematodes associated with grapevine degeneration in Michigan. Phytopathology. 1974;**64**:1174-1178

[44] Lammers AH, Allison RF, Ramsdell DC. Cloning and sequencing of peach rosette mosaic virus RNA1. Virus Research. 1999;**65**:57-73

[45] Allen W, Schagen JV, Ebsary B. Comparative transmission of the peach rosette mosaic virus by Ontario populations of Longidorus diadecturus and *Xiphinema americanum* (Nematoda: Longidoridae). Canadian Journal of Plant Pathology. 1984;**6**:29-32

[46] Sanfaçon H, Iwanami T, Karasev A, Van der Vlugt R, Wellink J, Wetzel T, et al. Family Secoviridae. In: Virus Taxonomy. Amsterdam, The Netherlands: Elsevier; 2011. pp. 881-900

[47] Fulton RW. Apple Mosaic Virus. CMI/AAB Descriptions of Plant Viruses No. 83. Wellesbourne, UK: Association of Applied Biologists; 1972

[48] Baumann G, Casper R, Converse RH. The occurrence of apple mosaic virus in red and black raspberry and in blackberry cultivars. In: ISHS Acta Horticulturae No. 129.1. III International Symposium on

Small Fruit Virus Diseases. Vancouver, B.C., Canada; 1982. pp. 13-20. DOI: 10.17660/ActaHortic.1982.129.3

[49] Baumann G. Virus infection in cropping and nursery plantations of raspberry and blackberry. I.Aphid-transmitted viruses. Erwerbsobstbau. 1984;**26**:188-192

[50] Converse RH, Casper R. A member of the apple mosaic virus group occurring in Rubus. In: Acta Horticulturae No. 44. 1975. p. 63

[51] Smith KM, Markham R. Two new viruses affecting tobacco and other plants. Phytopathology. 1944;**34**:324-329

[52] Mossop DW, Fry PR, Young BR. New plant disease records in New Zealand: Arabis mosaic virus in celery, lettuce and Chinese cabbage; tomato spotted wilt virus in celery. New Zealand Journal of Agricultural Research. 1983;**26**:257-259

[53] Murant AF. Arabis mosaic virus. CMI/AAB Descriptions of Plant Viruses No. 16. Wellesbourne, UK: Association of Applied Biologists; 1970

[54] Brown DJF. The transmission of two strains of Arabis mosaic virus by populations of *Xiphinema diversicaudatum* (Nematoda: Dorylaimoidea) from ten countries. Revue de Nématologie. 1986;**9**:83-87

[55] Trudgill DL, Brown DJF, McNamara DG. Methods and criteria for assessing the transmission of plant viruses by longidorid nematodes. Revue de Nématologie. 1983;**6**:133-141

[56] Stace-Smith R. Virus and virus-like diseases of *Rubus* (raspberry and blackberry). In: Converse RH, editor. Virus Diseases of Small Fruits. Washington, D.C: U.S. Government Printing Office; 1987

[57] Stace-Smith R. Studies on Rubus virus diseases in British Columbia. II. Black raspberry necrosis. Canadian Journal of Botany. 1955;**33**:314-322

[58] Martin RR, MacFarlane S, Sabanadzovic S, et al. Viruses and virus diseases of Rubus. Plant Disease. 2013;**97**:168-182

[59] Jones AT, Wood GA. The virus status of raspberries (*Rubus idaeus* L.) in New Zealand. N.Z. Journal of Agricultural Research. 1979;**22**:173-182

[60] Halgren A, Tzanetakis IE, Martin RR. Identification, characterization, and detection of black raspberry necrosis virus. Phytopathology. 2007;**97**(1):44-50. DOI: 10.1094/phyto-97-0044 Available from: https://www.dpvweb.net/dpv/showdpv/?dpvno=333

[61] Stace-Smith R, Jones A. Black raspberry necrosis. In: Converse RH, editor. Virus Diseases of Small Fruits. Washington: US Government Printing Office; 1987

[62] Thompson JR, Perry KL, De Jong W. A new potato virus in a new lineage of picorna-like viruses. Archives of Virology. 2004;**149**:2141-2154

[63] Jones AT, Badenoch S. New virus in raspberry from Canada. In: Scottish Horticultural Research Institute Annual Report for 1980. Dundee Scotland; 1981. p. 100

[64] Francki RIB, Mossop DW, Hatta T. Cucumber mosaic virus. In: Descriptions of Plant Viruses No. 213. Kew, Surrey, England: Commonwealth Mycological Institute and the Association of Applied Biologists; 1979. p. 6

[65] Kennedy JS, Day MF, Eastop VF. A conspectus of aphids as vectors of plant viruses. Commonwealth Institute of

Entomology. London: Eastern Press Ltd.; 1962. p. 114

[66] Harrison BD. Cucumber mosaic virus in raspberry. Plant Pathology. 1958;**7**:109-111

[67] Jones AT. An isolate of cucumber mosaic virus from *Rubus phoenicolasius* maxim. Plant Pathology. 1976;**25**:137-140

[68] Gordejchuk OG, Krylov AV, Krylova NV, Samonina IN. Virus diseases of berry crops in the soviet Far East. I. Identification of some mechanically transmitted viruses, detected in Primorye territory. Zentralblatt für Bakteriologie, Parasitenkunde, Infektionskrankheiten und Hygiene, Abteilung II. 1977;**132**:686-707

[69] Derrick KS, Beretta MJ, Barthe GA. Detection of an *Idaeovirus* in citrus by subtraction hybridization. (Abstr.). Phytopathology. 2005;**95**:S24

[70] Jones AT, Murant AF, Jennings DL, Wood GA. Association of raspberry bushy dwarf virus with raspberry yellows disease; reaction of *Rubus* species and cultivars, and the inheritance of resistance. The Annals of Applied Biology. 1982;**100**:135-147

[71] Knight VH, Barbara DJ. A review of raspberry bushy dwarf virus at HRI-East Malling and the situation on sample of commercial holdings in England in 1995 and 1996. Acta Horticulturae. 1999;**505**:263-271

[72] Jones AT. The effects of black raspberry necrosis and raspberry bushy dwarf viruses in Lloyd George raspberry and their involvement in raspberry bushy dwarf disease. Journal of Horticultural Sciences. 1979;**54**:267-272

[73] Jelkmann W, Martin RR. Complementary DNA probes generated from double-stranded RNAs of a *Rubus* virus provide the potential for rapid *in vitro* detection. Acta Horticulturae. 1988;**236**:103-109

[74] McGavin WJ, Mitchell C, Cock PJA, Wright KM, MacFarlane SA. Raspberry leaf blotch virus, a putative new member of the genus *Emaravirus*, encodes a novel genomic RNA. The Journal of General Virology. 2012;**93**:430-437

[75] Jones AT, Gordon SC, Jennings DL. A leaf-blotch disorder of tayberry associated with the leaf and bud mite (*Phyllocoptes gracilis*) and some effects of three aphid-borne viruses. Journal of Horticultural Sciences. 1984;**59**:523-528

[76] Bennett CW. Further observations and experiments on the curl disease of raspberries. Phytopathology. 1930;**20**:787-802

[77] Bolton AT. Spread of raspberry leaf curl virus. Canadian Journal of Plant Science. 1970;**50**:667-671

[78] Ebel R, Schnabel A, Reustle GM, Krczal G, Wetzel T. Complete nucleotide sequence of an isolate of the nepovirus raspberry ringspot virus from grapevine. Virus Research. 2003;**97**:141-144

[79] Cadman CH. Studies on the etiology and mode of spread of Scottish raspberry leaf curl disease. Journal of Horticultural Sciences. 1956;**31**:111-118

[80] Murant AF. Raspberry ringspot virus. CMI/AAB Descriptions of Plant Viruses No. 198. Wellesbourne, UK: Association of Applied Biologists; 1978

[81] McGavin WJ, Cock PJA, MacFarlane SA. Partial sequence and RT-PCR diagnostic test for the plant rhabdovirus raspberry vein chlorosis virus. Plant Pathology. 2011;**60**:462-467

[82] Jones AT, McGavin WJ, Geering ADW, Lockhart BEL. Identification of Rubus

yellow net virus as a distinct badnavirus and its detection by PCR in *Rubus* species and in aphids. The Annals of Applied Biology. 2002;**141**:1-10

[83] Stace-Smith R, Jones AT. Rubus yellow net virus. In: Converse RH, editor. Virus Diseases of Small Fruits. No. 631. Washington, D.C.: USDA Agriculture Handbook; 1987. pp. 175-178

[84] McGavin WJ, MacFarlane SA. *Rubus chlorotic mottle* virus, a new sobemovirus infecting raspberry and bramble. Virus Research. 2009;**139**:10-13

[85] Kado CI. CMI/AAB Descr. In: Plant Viruses – Sowbane Mosaic Virus. No. 64. 1971. p. 4

[86] Madhusudan T, Govindu HC. Identification of virus diseases of tobacco in South India. Indian Journal of Virology. 1985;**1**:163-173

[87] Von Bargen S, Langer J, Robel J, Rumbou A, Büttner C. Complete nucleotide sequence of *cherry leaf role virus* (CLRV), a subgroup C nepovirus. Virus Research. 2012;**163**:678-683

[88] Cropley R, Tomlinson JA. Cherry leaf roll virus. CMI/AAB Descriptions of Plant Viruses No. 80. Wellesbourne, UK: Association of Applied Biologists; 1971

[89] Jones AT, Wood GA. The occurrence of cherry leaf roll virus in red raspberry in New Zealand. Plant Disease Report. 1978;**62**:835-838

[90] Al-Chaabi S, Ismaeil F. First report of the cherry strain of cherry leaf roll virus on walnut and pecan trees in Syria. Journal of Plant Pathology. 2009;**91**:502

[91] Sanfaçon H, Iwanami T, Karasev AV, van der Vlugt R, Wellink J, Wetzel T, et al. Secoviridae. In: King AMQ, Adams MJ, Carstens EB, Lefkowitz EJ, editors. Virus Taxonomy, 9th Report of the International Committee on Taxonomy of Viruses. San Diego, CA, USA: Academic Press; 2012

[92] Murant AF. CMI/AAB Descr. In: Plant Viruses – Strawberry Latent Ringspot Virus. Descript of Plant Viruses 126. Kew, Surrey, England; 1974. 4 p

[93] Rott ME, Tremaine JH, Rochon DM. Nucleotide sequence of tomato ringspot virus RNA 2. The Journal of General Virology. 1991;**72**:1505-1514

[94] Rott ME, Gilchrist A, Lee L, Rochon D. Nucleotide sequence of tomato ringspot virus RNA 1. The Journal of General Virology. 1995;**76**:465-473

[95] Fromme FD, Wingard SA, Priode CN. Ringspot of tobacco: An infectious disease of unknown cause. Phytopathology. 1927;**17**:321-328

[96] Rush MC, Gooding GV, Ellis DV. Wild *Rubus* spp. as natural hosts for tobacco ringspot virus. Phytopathology. 1968;**58**:1065

[97] Stace-Smith R. CMI/AAB Descr. In: Plant Viruses - Tomato ringspot virus. Descript of Plant Virus. No. 18. Kew, Surrey, England; 1970. 4 p

[98] Anonymous. EPPO data sheet on quarantine organisms no. 102: Tomato ringspot virus. OEPP/EPPO Bulletin. 1982;**12**:6

[99] Pinkerton JN, Martin RR. Management of *tomato ringspot virus* in red raspberry with crop rotation. International Journal of Fruit Science. 2005;**5**:55-67

[100] Freeman JA, Stace-Smith R. Effects of tomato ringspot virus on plant and fruit development of raspberries. Canadian Journal of Plant Science. 1968;**48**(1):25-29

[101] Stace-Smith R, Converse RH. Tomato ringspot virus in Rubus. In: Converse RH, editor. Virus Diseases of Small Fruits. Vol. 631. Washington, USA: USDA, Agricultural Handbook; 1987. pp. 223-227

[102] Ellis P, Stace-Smith R, de Villiers G. Identification and geographic distribution of serotypes of potato virus Y. Plant Disease. 1997;**81**(5):481-484

[103] FAO. Pest risk analysis for quarantine pests, including analysis or environmental risks and living modified organisms. In: Int. Stand. Phytosanit. Meas. Number 11. Rome: Food Agric. Organ; 2004

[104] FAO. Framework for pest risk analysis. In: Int. Stand. Phytosanit. Meas. Number 2. Rome: Food and Agriculture Organization of the United Nations; 2007

Chapter 4

Management Strategies for Botrytis Rot in Strawberry

Abdul Rehman, Faizan Ali, Akhtar Hameed and Waqar Alam

Abstract

Strawberry is major horticultural fruit crop grown across the globe. The crop is basis of a multibillion dollars food product industry and a major employer of the global population. Gray mold disease caused by pathogen *Botrytis cinerea* is responsible for massive pre-harvest and post-harvest losses in the crop making it a major challenge for the overall health of the industry. Furthermore, infected fruits are equally disliked by commercial buyers and domestic consumer resulting in to major losses for the growers. Rising populations and climate change factors are crucial in this aspect as well, because they have a negative impact on the overall yield of crop and can also lead to rapid mutations in pathogen genome making it more resilient to multiple climatic conditions. In this chapter we are going to discuss fundamentals of this issue, basic biology of the pathogen followed by conventional and modern approaches for disease control and future perspective.

Keywords: strawberry, botrytis rot, pre-harvest, post-harvest losses, management

1. Introduction

Strawberry is a globally important fruit crop with significant financial value for growers across the globe. Currently, almost 400,000 hectares of land is cultivated with strawberry crop plant with estimated global market value of about 19 billion USD [1]. China and US are the biggest producers of the crop with US market worth about US$ 2.5 billion [2]. Most common commercial cultivar of the plant has originated some 300 years before in response to hybridization between four different global cultivars (*F. viridis, F. iinumae, F. nippiconica* and *F. vesca*) [3]. Current domesticated crop is an allo-octoploid crop (2n = 8x = 56) and its overall genomic complexity has often led to the use of diploid relatives as model crops to better understand the genomic complexion of crop.

Commercial cultivars of the crop found their origin about 3 centuries ago in accidental hybridization of *Fragaria chiloensis* and *Fragaria virginiana* cultivars. Evidences has shown that Thomas Knight made some of the earliest efforts for crossing and breeding of strawberry cultivars in his personal gardens during 1817 in Britain [4]. Similarly, growers from north American region also began making efforts for newer and better cultivars of the crop, and these efforts played an important in overall technological production advancement and breeding efforts of the crop over the next 2 centuries. Strawberries and other berry fruits (i.e. blueberries, raspberries, blackberries, etc.) are well known for their high crop nutritional profile and are often added in to diet plan for patients of various cardo-vascular diseases [5]. Strawberry is among the group of so-called super fruit group that often tend to enhance overall body metabolism and bio-chemical setup with an additional supply of minerals and anti-oxidants. According to an estimation made by the UNFAO, overall production of strawberry plant has increased by more than 80% over the last two decades with nearly 3/4th of the produced berries being consumed as fresh products while remaining being taken in by industries as processed foods.

In general strawberry are perennial plants with an herbaceous growth pattern, and denser leave system. It produces a complex fruit called the achenes which are single-seed fruit, and the receptacles that are similar to floral meristem tissues [6]. Strawberries are source of various vitamins and minerals that are essential for proper functioning of human body. Strawberry fruit also contain good amount of dietary fibers that help in maintenance of blood sugar levels, and healthy fatty acids helps in metabolism stability.

Strawberry crop is infected by a number of pathogens including Fungal, bacterial, viral and nematode based infections. Among these most damaging of all are fungal pathogens that result in major losses for the crop [7]. Fungal pathogens are capable of damaging all parts of the plant including stem, root, leaf and fruit, in addition to this these pathogens are responsible for causing on field as well as post-harvest losses of the crop. Among all the fungal pathogens gray mold or the botrytis disease is the most damaging one caused by ascomycetes *Botrytis cinerea,* as it is responsible for significant on field yield losses as well as post-harvest losses [8]. The pathogen has severe economic implications for the growers and causes senescing of plant fruit and vegetative tissues. Disease prevalence is more dominant under wet, humid conditions, with some infection leading up to 80% losses (**Table 1**).

Ploidy	Species	Indigenous Area	Fruit Characters
Diploid	*Fragaria nubicola*	Himalayas	Bright red appearance with raised seed
	Fragaria viridis	Central Europe/ Eastern Europe	Small and firm fruit with Pink-red appearance
	Fragaria daltoniana	Himalayas	Elongated and bright red fruit with no taste
	Fragaria iinumae	Japan	Ovoid in appearance with sunken achenes
	Fragaria vesca	Japan	Bright red appearance with raised seed
Tetraploid	*Fragaria mounpinensis*	China—Tibet region	Small fruit with similarity to *F. nilgerrensis*
	Fragaria orientalis	Korea and Siberia regions	Soft fruit, with slight aroma and sunken seed
Hexaploid	*Fragaria moschata*	North Europe	Light to Dark purplish red color, ovoid shape, raised achenes

Ploidy	Species	Indigenous Area	Fruit Characters
Octoploid	*Fragaria virginiana*	North America	Double than the size of *F. vesca* with soft fruit and light to deep red in color appearance
	Fragaria chiloensis	North America, Chile, Argentina	Red brownish furit color with white flesh
	Fragaria ovalis	Western coast to Rocky Mountains	Round small sized fruit with pink color
	Fragaria inturupensis	North Japan	Oval shaped fruit

Table 1.
Species diversity of strawberry plant (Fragaria spp.).

2. Botanical aspects of strawberry cultivars

One of the earliest efforts made to answer several mysteries of strawberry plant and discussing its botanical efforts were made by French researchers Antoine Nicolas Duchesne and Bernard de Jussieu, and their findings were later on published in their book during 1766 in France [9]. In their book Duchesne discussed basic aspects of strawberry plant and concluded that the current commercial cultivar of the crop is hybrid made in North America from two South American cultivars of the crop (i.e. *F. virginiana* and *F. chiloensis*) [10]. The genus Fragaria belongs to family Rosaceae that also contains a large number of commercially significant fruit crops and ornamental plants in it. Fragaria genus is added in to sub-family group Rosoideae that has about 20 difefernet species of the plant added in to it, with some of those being diploid, tetraploid, hexaploid and octaploid varieties [11]. The sub-family also consists a number of wild relatives of the crop such as, *Fragaria nubicola* and *Fragaria vesca* which are diploid wild relatives of commercial strawberries. Strawberry plants are normally perennial herbaceous plant that have trifoliate, membranous leaves, pitted leaflets and whitish flowers. The petals are normally obovate, short, scrabbled and numerous distinct pistils borne on elevated convex receptacles. These receptacles later on provide space for berry like fruits to grow on, with minute seed like achenes formed over the fruit [12].

An important aim of modern breeding programs is to improve growth potential of the plant as well as betterment of nutritional profile. These features are controlled and manipulated by genetic and epigenetic elements of the plant with some changes being permanently added in to the plant while other being a temporary addition in to their nutritional profile due to influence of favorable conditions [13]. Several quality traits are greatly influenced in strawberry plants owing to genetic basis as well as climatic factors associated with growing practices of plant, these traits include size of the fruit, it shape, color, rigidity of the fruit pulp and overall color formation on fruit surface. Over the time growing awareness among consumers is pushing people worldwide to add more proportion of food in their daily intake for healthy and balanced diet. For this purpose, small berry fruits are a major contender owing to its higher concentration of numerous vitamins, minerals, phenolic compounds, flavonoids, iso-flavonoid etc. Owing to this major health benefit associated with the crop it has become an important target for researchers to identify cultivars with relatively higher concentration of beneficial bio-compounds and subject these varieties for breeding

Class	Group	Compound
Flavonoids	Anthocyanins	Cyanidin-3-glucoside
		Cyanidin-3-malonylglucoside
		Pelargonidin-3-galactoside
		Pelargonidin-3-arabinoside
		Pelargonidin-3-malyloglucoside
		Pelargonidin-3-acetylglucoside
	Flavonols	Quercetin-3-glucuronide
		Quercetin-glucoside
		Quercetin-glucuronide
		Kaempferol-glucunoride
Phenolic acids	Hydroxycinnamic acids Ellagitannins	Ellagitannin
		Bis-HHDP-glucose
		Methyl-EA-pentose conjugate
		Ellagic acid
		Sanguiin H-6

Table 2.
Biochemical profile of strawberry fruit.

programs. Strawberry fruit has a diverse and heathy nutritional profile, among the most important of this are phenolic compounds which play an important role in biochemical activities, growth and survival of plant under different growing conditions. Basic structure of phenolic compounds consists of one or more benzene ring in it with increased solubility owing to attached sugar molecules with it. Flavonoids, tannins, and phenolic acids are the most significant biochemicals of the category essential for optimum growth of plants and stable functioning of human body (**Table 2**) [14].

Phenolic compounds are produced in response to plant biochemical activity, they are essential for normal growth and survival of plants [15]. Evidence shows that these compounds are bio-actively important in humans as well owing to consumption of berry fruit. Several groups of phenolic compounds are important in this regard including flavonoids, hydrolysable tannins, phenolic acids, and condensed tannins. Anthocyanin is an important hydrophilic pigment associated with strawberry fruit and it has a varying color ranging from reddish to slight purple in color, this varying pigmentation in fruit is attributed to growing pH conditions of plant [16]. These anthocyanin compounds have a higher concentration of anti-oxidant compounds increasing overall antioxidant capacity of plant. Berry fruit also contains fair amount of carbohydrate molecules in it, including fructose, glucose and sucrose, also various organic acids such citric acid, malic acid etc. are all present in strawberry.

3. Disease pathogen

B. cinerea has a wide range of host and is capable of causing infection on hundreds of plant species having significant food and ornamental value [17]. Many horticultural crops (fruits and vegetables) have been reported to accumulate

DOI: http://dx.doi.org/10.5772/intechopen.1002225

millions in yield losses owing to *B. cinerea* pathogen infestation. The pathogen has a necrotrophic mode of nutrition that grows on host plant vegetative tissue ultimately causing death of plant. Pathogen spores are predominantly present in different growing locations that emerge out of infected plant parts which serve as primary inoculum for causing infection. Initially spore gain entrance inside plant system via several natural openings or wounds [18]. Although it has been reported to cause significant damage in mature, ripped plants, infection rates are observed to be lower in unripen plants. Often in unripe plants pathogen has been reported to cause slow infection either in of delayed germination of conidia, as symptomless endophytic infection or colonization of abscising plant organs where growth arrests. Often under mild infectious conditions, pathogen enters a small asymptomatic phase during initial stages of the cycle. Followed by this pathogen initiates a more severe necrotrophic cycle in ripening plant tissues that leads to cellular breakdown and decay of these tissues [19].

Owing to modern sequencing technologies and OMIC approaches, more information is now available regarding bio-chemical basis of the plant as well as the corresponding pathogen species [20]. Reference genome analysis for the botrytis pathogen has shown that pathogen produces a variety of virulence factor which promote overall susceptibility of the pathogen [21]. During initial stages of the infection the pathogen is known to secrete a variety of effector proteins molecules and sRNA's that enable suppression of cell death under stress and restrict immuno-response of the plant [22]. This provides favorable condition for plant to further grow in plant tissues and gradually establish infection in plants. Establishing fungal hyphae also secrets dicer like protein molecules DCL1 and DCL2 which penetrates in to the host cell and inhibit its RNAi mechanism limiting host defense ability in response to the infection [23]. Several pathogen enzymes including toxins and reactive oxygen species (ROS) producing enzymes cause immediate death of host cell. Another key activity involves synthesis of oxalic acid that lowers pH level of host cell, creating a suitable environment for enhanced fungal activity. Pathogen produces several enzymes including pectinases, protease and laccase that result in cellular degradation [24]. Pathogen also cause accumulation of Ca^{2+} which leads to lower structural integrity of pectin in cell wall and restrict callose deposition in cellular structures. Furthermore, pathogen also secrets cell wall degrading enzymes which causes cell wall to loosen and lysis of cell. Fungal pathogen also secrets a variety of hormonal biochemical which interfere with normal cellular functioning of host cell.

4. Pathogen infection mechanism

Infection in strawberry plants is caused by invading fungal pathogen, in primary infection invading pathogen hyphae tend to infect the host plant specially during the flowering stage, grow in to receptacle during flowering stage. Pathogen often tend to overwinter by forming sclerotia or by infecting surrounding plants with mycelial extension. Pathogen growth is restricted after initial infestation of unripe receptacle and a symptomless quiescent phase occurs. Estimates has shown that inhibition of earlier infection in unripe plants is nearly 50% in comparison to just 8% in ripen plant fruit. This is attributed to the presence of proanthocyanins in unripe fruit which restrict the activity of fungal enzymes necessary for rapid infection of plant [25]. Similarly, anthocyanin content has also been associated with lower infection in early growth stages of plant.

In secondary infection the source of conidia is diverse and it initiate necrotrophic phase without any quiescence. Secondary infection takes place at a rapid pace with initial symptoms appearing within 16 hours of primary infection and significant biomass appearing in almost 48 hours of initial infection [26]. Early responses of strawberries to infection include higher expression of the defense genes FaPGIP and FaChi 2–1 (Class II Chitinase), whereas lower expression of the reference gene DNA Binding Protein – FaDBP indicates extensive cell death induced by *B. cinerea* at late stages of infection [27].

Research evidence suggests that ripening process tend to increase the overall susceptibility of strawberry fruit to Botryits cinerea infection. In unripe form berry fruit is resistant to fungal infection as it causes pathogen quiescence. Initiation of ripening process tend to increase it susceptibility, this is due to sveral transcriptional regulator molecules associated with ripening process that increase disease risk of strawberry [28]. OMIC analysis has shown that significant changes takes place during trnastion from large green to white phase; this includes alteration in cell wall composition, biochemical profile, plant hormones, pigmentation, carbohydrate metabolism and antioxidant profile [28]. This result in relatively lower oxidative phosphorylation during ripening ultimately disrupting a variety of biochemical and physiological processes. Research has shown strong evidence suggesting that pathogen infection interferes with biosynthetic mechanism of plant hormone synthesis including; ethylene, ABA, etc., that tend to act as virulence factor by promoting fruit ripening and senescence process (**Figure 1**) [29].

Pathogen also tend to introduce several physiological and cellular changes in the crop for facilitating infection process. Cell wall degradation assist higher infection severity as the physical barriers to pathogen invasion are weakened. In berry fruit, cell wall begins to disassemble as the fruit enters in to ripening phase, this solubilization of cell wall increases overall available sugar contents as polysaccharides continue to breakdown. Down regulation of pectin lyase (PL) gene has shown a higher degree

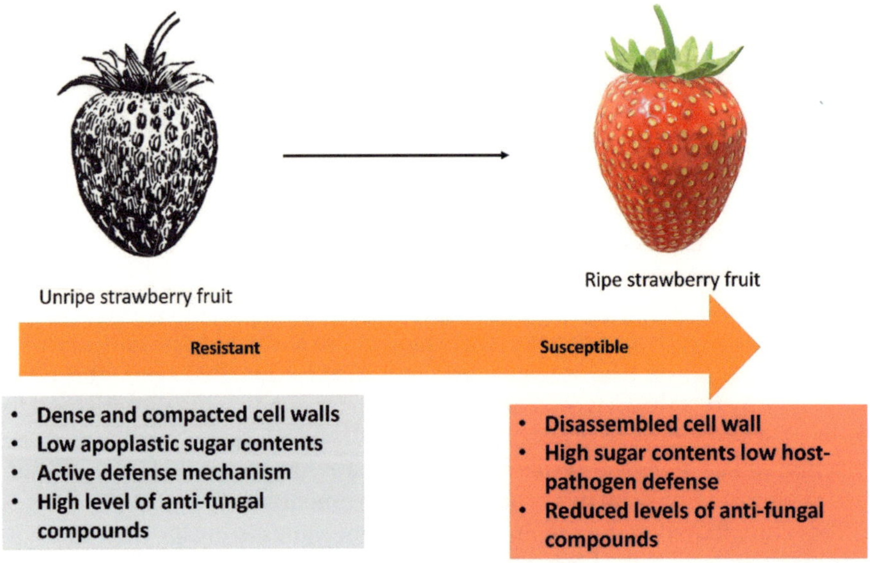

Figure 1.
Influence of ripening phenomenon on overall susceptibility of strawberry fruit to B. cinerea infection.

of fruit firmness during ripening and a lowering overall severity of disease. Invading pathogen targets polysaccharide compounds in cell wall by secretion of a variety of cell wall degrading enzyme, pathogenicity gene such as Bcpg (endopolygalacturonases gene) gene plays an important role in infection prevalence [27, 30]. Similarly, cuticle is another major barrier to prevent pathogen invasion, and this barrier get compromised during ripening phase as expanding fruit tend to create small cracks on fruit surface and relatively thinner cuticle. In addition to this pathogen secret cuticle degrading enzyme which further increase overall penetration and infection causing ability of pathogen [28].

5. Management strategies for botrytis rot in strawberry

Owing to the overall quantitative and qualitative importance of the strawberry fruit crop, management of pathogen is a crucial task in this regard. Historically, little information was available regarding pathogen survival, overwintering in other crops, mode of action and optimum growing conditions that limited the potential of control measures. Owing to this lack of information regarding disease spread the disease was mainly managed via conventional or cultural practices. During the late 19th to early 20th century discovery of pathogen species, formulation of new chemical compounds and the green revolution enabled application of synthetic chemicals as well as certain biocontrol agents for disease management. Recent progress in the availability of OMIC's data owing to rapid progress in sequencing technologies are now enabling researchers to introduce resistance capabilities in to the genome of plant to overcome the invading pathogen. Latest genetic engineering technologies like the CRISPR/CAS technology has enabled knocking-out of susceptibility genes in plant and virulence genes in pathogen, making it difficult for disease to affect the plant. Also, massive data sets regarding climatic conditions has added in to our better understating of optimum growth conditions of both the plant and pathogen which can be used for managing infections in post-harvest and controlled growing environments. Following are some of the key management strategies used for managing Botrytis rot infection.

5.1 Conventional management techniques

Historically botrytis infection has been contained by adopting certain management strategies focused on constraining inoculum buildup of pathogen. Similarly, second most common approach used involves the practice of keeping pathogen away from the fruit by avoiding fruit contact with soil, infected plant parts or contaminated farm equipment. It is important to avoid contact of fruit with soil surface as certain conidia are present on the surface of soil and tend to germinate quickly upon connection with fruit and under relatively higher soil moisture [31]. Owing to pathogen dependence on higher soil moisture content it is also important to better manage the field irrigation system. In this regard, drip irrigation system, micro-sprinklers and smart sensors-based irrigation system plays an important role in maintaining stable soil moisture as well as restricting over irrigation of field to avoid prevalence of conidial germination [32]. Shady environment also tends to aid fungal infection and for this purpose proper thinning and pruning of trees tend to reduce infection initiation probability. Furthermore, application of moderate to lower dosage of nitrogen fertilizers is recommended during infection season to avoid excessive shoots and leaves growth. Similar to canopy issue, close spacing in plants also tend to promote infection and can be avoided with appropriate spacing among fruit plants. Another common practice for reducing fungal infestation is to grow plant under tunnels, in this case a relatively lower level of air inoculum is able to invade host which in turn reduces overall rate of infection. Although beneficial, tunnel-based approaches are challenging as well, the tend to increase risk of powdery mildew infection and create issues with fruit harvesting process [33]. Conclusively, the conventional approaches have shown significant results in the past and still plays a prominent role in management of disease.

5.2 Fungicide/chemical based control strategies

Demand for increase of crop production has always been a major challenge for human race, in an effort to satisfy increasing food demands of rising global population. This effort has often faced plant disease as a major challenge that tend to restrict fruitful outcomes of such efforts. In this regard application of various pesticide or insecticide chemicals has been utilized as a major source to restrict the pathogen activity and ensure higher crop yields [34]. Like many important plant diseases (such as powdery mildew disease, downy mildew, rice blast disease etc.) botrytis rot of strawberry is also manage heavily these days via synthetic pesticide chemicals. In order to secure effective outcomes from pesticidal application it is important to take in account for number of applications and time span for applying chemicals to diseased or susceptible plant [35]. furthermore, use of multiple fungicide is also important to control such pathogens that requires multi-mode of actions for growth inhibition. Botrytis rot has been heavily managed by the use of such chemicals with fungicides from the FRAC (Fungicide resistance action committee) groups and captain being most prominent group of chemicals overall [34]. Recently resistance development to fungicide application in pathogen has been a major challenge in plant disease management owing to changing resistance profile of invading pathogens in a single growing season (**Table 3**).

Efforts have been made recently in order to develop resistance profile of several *B. cinerea* isolates, a study conducted in the Louisiana state of USA involved resistance profiling of nearly 13 fungal isolates with each of them showing lower to moderate level resistance against FRAC 1 pesticidal chemicals. These isolates were later on further tested against FRAC 2 type of pesticide chemicals with a few isolates showing some degree of resistance to them. Another study including assessment of more than 1800 *B. cinerea*

FRAC Chemical	Target site	Target action	Resistance risk profile
Dicarboximides	Hisitidine kinase	Signal transduction	Medium to high risk
Succinate dehydrogenase inhibitor	Succinate dehydrogenase	Respiration activity	Medium to high risk
Benzimidazole	Tubulin assembly	Cytoskeleton	High rsik
Phenylpyrroles	Hisitidine kinase	Signal transduction	Low to medium risk
Sterol bio-synthesis inhibitor	Keto reductase in C4 de-methylation	Inhibition of sterol synthesis	Low to medium risk
Anilinopyrimidines	Methionine synthesis	Protein synthesis	Medium risk

Table 3.
List of synthetic fungicide chemicals to control botrytis disease.

isolates was carried out in 2015, isolates from 10 different American states were subjected to resistance profiling, results indicated that multiple isoaltes have developed resistance against all single action sites of FRAC group chemicals [36]. A positive correlation was also observed between overexpression of efflux transporters, modification of fungicide target site and increased resistance level of pathogen against chemicals. *B. cinerea* also possess a diverse range of transposable elements in its genome, along with heterokaryosis and sexual reproduction process that enables pathogen to gain multiple resistance related mutations in a single growing season. This increased resistance in fungicide action sites indicates a prevailing nascent demand for more innovative approaches in disease management strategy. Efforts are now under way to produce new type of fungicide chemicals that will interfere with pathogen growth patterns at RNA level and inhibition of transcription-translation phases in plants, enabling targeted growth reduction and containment of pathogen infestation [37]. These approaches are in early development phase and far from large scale commercial availability. Meanwhile, mixed application and rotation of different fungicide chemicals can assist better management of disease along with avoiding risk of resistance development.

5.3 Biological control approaches for *B. cinerea*

Although synthetic chemicals have been utilized heavily for disease containment, many of the negative aspect of aforementioned approach has raised questions regarding its effectiveness. For example, many copper based or other type of pesticide chemicals are causing respiratory, dermatological issues as well as heavy metal contamination in agricultural fields [38]. Therefore, evaluation and application of other eco-friendly sustainable approaches has become a priority now. In this regard use of biocontrol microbes has been under extensive studies to make commercially viable eco-friendly products out of it. Several microbial species including *Bacillus subtilis, Colletotrichum gloeosporioides, Epicoccum purpurascens, Gliocladum roseum, Penicillium sp., Trichoderma spp.* have been tested to contain botrytis infection in controlled conditions [39]. Earlier studies have indicated that in many cases bio-control agents tend to reduce disease severity much higher than the synthetic chemicals, with up to 90% disease reduction in stamens and more than 75% reduction in fruits.

Bio-control agents utilizes multiple mode of actions to contain pathogenic microbe growth, this includes competition for primary growth nutrients, secretion of several growth inhibiting bio-molecules (such as anti-biotics), and influencing host plant biomechanism to produce pathogen growth restricting molecule such as chitinases and peroxidases, as well as certain PR proteins [40]. A combination of bio-control microbes is recommended to for application to infected plant so that pathogen growth contained via multi-mode of action strategy. Also, use of certain bio-control volatile compounds and extracts from these microbes has shown some potential to resolve pathogenicity issue [41]. Overall microbes and microbes-based bio-compounds have shown significant growth inhibiting properties, but their large-scale commercial application remains under question due to high degree of pathogen diversity at sub-species and variant levels, as well as overall commercial cost that is needed to incurred in order to produce commercial scale quantity of microbial compounds for disease management.

5.4 Genomic approaches for disease management

Over the last three decades huge number of genomic information data sets have been generated related to various crop plants and disease pathogens. These data sets have enabled a better understanding related to genes responsible for susceptibility in plants and utilization of this information for better management of disease. Earlier efforts were mainly the transgenic approaches that involved Agrobacterium mediated cellular transformation for disease related traits. Several factors affecting transformation efficiency in strawberry have been determined, that the leaf discs from in vitro cultures proliferating in the presence of 2.21 μM kinetin were the best explant for transformation. Furthermore, it was observed that the transformation efficiency for antibiotic-sensitive *F. vesca* and *Fragaria semperflorens* could be improved by using antibiotic carbenicillin for selection and suitable Agrobacterium strain. They achieved optimal transformation efficiency (15%) by the appropriate use of explant type and age, leaf-disc orientation, inoculation time, and phenolic compounds for bacterial virulence induction [42]. Incubation of *A. tumefaciens* with acetosyringone and indole acetic acid, age of explant, pre-culture, and pre-selection on antibiotic-containing medium were the other key factors to affect the transformation efficiency in strawberry.

Studies have shown significant results obtained by transgenic techniques for management of botrytis rot in strawberry. Schestibratov and Dolgov (2005) introduced thaumatin II protein via Agrobacterium-mediated transformation in multiple strawberry plants, with transformed platns with higher level of thau II expressed proteins exhibiting resistance against botrytis infection [42]. Vellicce et al. focused on strengthening of cellular defense mechanism in plant by transforming such genes in to plants. Two genes were targeted for this purpose; ch5B gene responsible for chitinase synthesis in plant and gln2 gene encoding for glucanase. Ch5B gene is responsible for degradation of cell-wall structures in pathogen resulting in to reduced growth of the pathogen and lesser infection severity [43]. In addition to this, modern CRISPR/CAS based genome editing has also enabled better management of disease; studies have shown that CRISPR mediated knock-out of SIMAPK3 gene susceptible for gray mold has shown a higher degree of resistance to the disease [44]. This is mainly due to increase in production of certain defense related secondary metabolites as well as synthesis of reactive oxygen species (ROS). In another study targeted mutagenesis of FaPG1 gene was carried out,

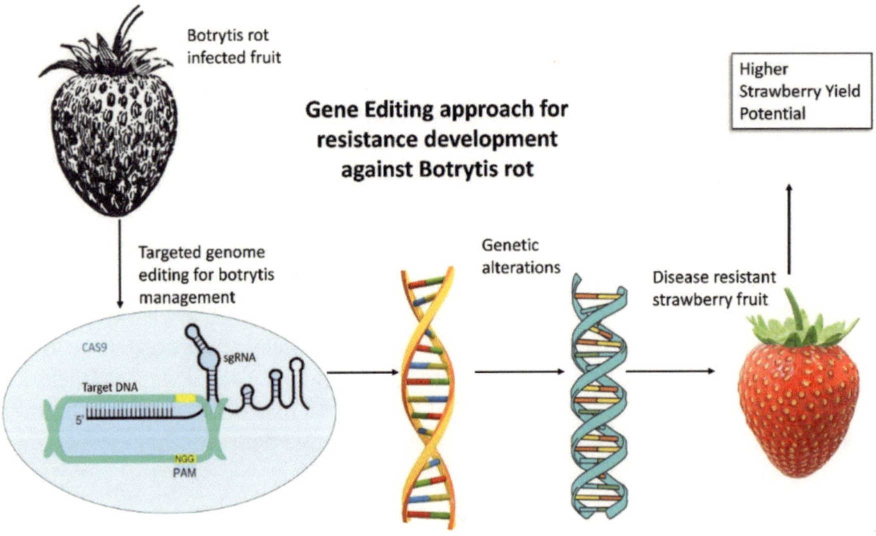

Figure 2.
CRISPR gene editing approach for resistance development in strawberry against B. cinerea infection.

this resulted in up to 70% increase in firmness of strawberry fruit, ultimately enhancing its shelf life and making less vulnerable to pathogen attacks (**Figure 2**) [45].

5.5 Post-harvest management of disease

The pathogen also causes major challenges during post-harvest handling of the crop ultimately resulting in major losses. Efforts are also made to alter storage conditions in a manner to reduce overall infection, berry fruit is stored immediately below 3 degree temperature which inhibits the chance of pathogen to regrow. Fruit is also kept in an environment that is higher in carbon dioxide concentration and lower in oxygen, so that metabolism of fruit is restricted [46]. Relative humidity during storage is usually kept around 85–90% to prevent dehydration of fruit, but limit fungal growth [47]. A novel approach for post-harvest management of this disease includes the application of protective coatings on fruit surface, that prevents loss of water from the fruit and at the same time restrict growth of fungus owing to presence of anti-fungal compounds. Application of chlorine mist in storage facilities also has the potential to reduce pathogen infestation and disease severity [48, 49]. More recent approaches includes use of ultraviolet radiation as a treatment method to kill any microbial spore that might be present on surface of harvested fruit or inside the storage facility.

6. Conclusion

Many aspects of botrytis disease development are still not properly understood, technological advancement will play an important role in better understanding genetic diversity of pathogen and various bio-chemical pathways crucial for pathogen cellular growth. Similarly, better understanding of plant defense mechanism and reasons associated with its failure will also assist in development of effective disease management strategies. Furthermore, current disease management needs to be re-evaluated to cope with increasing restrictions and lack of efficacy of fungicides. Investigations on biocontrol approaches and pre- and postharvest treatments are necessary to manage gray mold. Research related to study of such natural mutations taken place in strawberry that enabled better resistance against pathogen attack can also assist in this regard. Also, massive information regarding climatic data can be used for better studying of climatic factors supporting pathogen growth so that a disease forecasting model can be developed which can later be used for better management of disease.

Conflict of interest

The authors declare no conflict of interest.

Author details

Abdul Rehman[1*], Faizan Ali[1], Akhtar Hameed[2] and Waqar Alam[3]

1 Department of Plant Pathology, University of Agriculture Faisalabad, Pakistan

2 Department of Plant Pathology, Institute of Plant Protection, MNS University of Agriculture Multan, Pakistan

3 Department of Plant Pathology, University of Okara, Pakistan

*Address all correspondence to: rehman.abdul@uaf.edu.pk; arb041@gmail.com

References

[1] FAO STAT. Food and Agriculture Organization of the United Nations. FAOSTAT Database. Rome, Italy: FAO. Available from: http://Faostat3.Fao.Org/Home/E [Accessed: October 9, 2017]

[2] USDA (2016) Noncitrus Fruits and Nuts 2016 Summary. Available from: http://usda.mannlib.cornell.edu/usda/current/noncfruinu/noncfruinu-06-27-2017.pdf

[3] Basu A, Nguyen A, Betts NM, Crifs LTJJ. Strawberry as a functional food: An evidence-based review. Critical Reviews in Food Science and Nutrition. 2014;**54**(6):790-806

[4] Scherr KD. Genetic and Environmental Factors Influencing Strawberry Production, Pollination, and Pathogens. Michigan, US: Oakland University; 2021

[5] Afrin S, Gasparrini M, Forbes-Hernandez TY, Reboredo-Rodriguez P, Mezzetti B, Varela-López A, et al. Promising health benefits of the strawberry: A focus on clinical studies. Journal of Agricultural and Food Chemistry. 2016;**64**(22):4435-4449

[6] Hollender CA, Geretz AC, Slovin JP, Liu ZJP. Flower and early fruit development in a diploid strawberry. Fragaria vesca. 2012;**235**:1123-1139

[7] Garrido C, Carbú M, Fernández-Acero FJ, González-Rodríguez VE, Cantoral JMJGGG. New insights in the study of strawberry fungal pathogens. Genes, Genomes, Genetics. 2011;**5**(1):24-39

[8] Rhouma A, Hajji-Hedfi L, Ben Othmen S, Kumari Shah K, Matrood AA, Okon OG, et al. Strawberry Grey Mould, a devastating disease caused by the airborne fungal pathogen Botrytis cinerea. Molecular Plant Pathology. 2022;**50**(2):44-50

[9] Williams RL. Botanophilia in Eighteenth-Century France: The Spirit of the Enlightenment. Germany: Springer Science & Business Media; 2001

[10] Hummer KE, Bassil N, Njuguna WJWCRG. Fruits BRT. Fragaria. 2011;**7**:17-44

[11] Dirlewanger E, Denoyes-Rothan B, Yamamoto T, DJG C. Genomics tools across Rosaceae species. In: Genetics and Genomics of Rosaceae. Springer, New York: Springer Science & Business Media; 2009. pp. 539-561

[12] Fait A, Hanhineva K, Beleggia R, Dai N, Rogachev I, Nikiforova VJ, et al. Reconfiguration of the achene and receptacle metabolic networks during strawberry fruit development. Plant Physiology. 2008;**148**(2):730-750

[13] Cobb JN, DeClerck G, Greenberg A, Clark R, McCouch SJT, Genetics A. Next-generation phenotyping: Requirements and strategies for enhancing our understanding of genotype–phenotype relationships and its relevance to crop improvement. Theoretical and Applied Genetics. 2013;**126**:867-887

[14] Saxena M, Saxena J, Nema R, Singh D, AJJ G. Phytochemistry. Phytochemistry of Medicinal Plants. 2013;**1**(6):168-182

[15] Pratyusha S. Phenolic compounds in the plant development and defense: An overview. In: Plant Stress Physiology. London, UK: Intech Open; 2022

[16] Khoo HE, Azlan A, Tang ST, Lim SMJF. Research n, Anthocyanidins

and anthocyanins: Colored pigments as food, pharmaceutical ingredients, and the potential health benefits. Food & Nutrition Research. 2017;**61**(1):1361779

[17] Elad Y, Williamson B, Tudzynski P, Delen N. Botrytis spp. and diseases they cause in agricultural systems–an introduction. In: Botrytis: Biology, Pathology and Control. Dordrecht: Springer; 2007. pp. 1-8

[18] Carisse O. Epidemiology and aerobiology of botrytis spp. In: Botrytis – The Fungus, the Pathogen and its Management in Agricultural Systems. Cham: Springer; 2016. pp. 127-148

[19] Molly FMD, Grant-Downton R. Botrytis-biology, detection and quantification. In: Fillinger S, Elad Y, editors. Botrytis – The Fungus, the Pathogen and its Management in Agricultural Systems. Cham: Springer; 2016

[20] Afridi MS, Ali S, Salam A, César Terra W, Hafeez A, Ali B, et al. Plant microbiome engineering: Hopes or hypes. Biology. 2022;**11**(12):1782

[21] González C, Brito N, Sharon A. Infection process and fungal virulence factors. In: Fillinger S, Elad Y, editors. Botrytis – The Fungus, the Pathogen and its Management in Agricultural Systems. Cham: Springer; 2016

[22] Van Kan JA, Stassen JH, Mosbach A, Van Der Lee TA, Faino L, Farmer AD, et al. A gapless genome sequence of the fungus Botrytis cinerea. Molecular Plant Pathology. 2017;**18**(1):75-89

[23] Weiberg A, Wang M, Lin F-M, Zhao H, Zhang Z, Kaloshian I, et al. Fungal small RNAs suppress plant immunity by hijacking host RNA interference pathways. Science. 2013;**342**(6154):118-123

[24] Schumacher J. Signal transduction cascades regulating differentiation and virulence in Botrytis cinerea. In: Fillinger S, Elad Y, editors. Botrytis – The Fungus, the Pathogen and its Management in Agricultural Systems. Cham: Springer; 2016

[25] van Baarlen P, Legendre L, van Kan JAL. Plant Defence compounds against botrytis infection. In: Elad Y, Williamson B, Tudzynski P, Delen N, editors. Botrytis: Biology, Pathology and Control. Dordrecht: Springer; 2007

[26] Harshman JM, Jurick WM, Lewers KS, Wang SY, Walsh CSJH. Resistance to Botrytis cinerea and quality characteristics during storage of raspberry genotypes. Hort Science. 2014;**49**(3):311-319

[27] Mehli L, Kjellsen TD, Dewey FM, Hietala AMJNP. A case study from the interaction of strawberry and Botrytis cinerea highlights the benefits of comonitoring both partners at genomic and mRNA level. New Phytologist. 2005;**168**(2):465-474

[28] Blanco-Ulate B, Labavitch JM, Vincenti E, Powell ALT, Cantu D. Hitting the wall: Plant cell walls during Botrytis cinerea infections. In: Fillinger S, Elad Y, editors. Botrytis – The Fungus, the Pathogen and its Management in Agricultural Systems. Cham: Springer; 2016

[29] Cantu D, Blanco-Ulate B, Yang L, Labavitch JM, Bennett AB, Powell ALJPP. Ripening-regulated susceptibility of tomato fruit to Botrytis cinerea requires NOR but not RIN or ethylene. Plant Physiology. 2009;**150**(3):1434-1449

[30] Rosli H, Civello P, Martínez GJPP. Changes in cell wall composition of three Fragaria x ananassa cultivars with different softening rate during ripening.

Plant Physiology and Biochemistry. 2004;**42**(10):823-831

[31] Dara SK, Sandoval-Solis S, Peck DJAS. Improving strawberry irrigation with micro-sprinklers and their impact on pest management. Agricultural Sciences. 2016;7(12):859-868

[32] Legard D, MacKenzie S, Mertely J, Chandler C, NJPD P. Development of a reduced use fungicide program for control of botrytis fruit rot on annual winter strawberry. Plant Disease. 2005;**89**(12):1353-1358

[33] Xiao C, Chandler C, Price J, Duval J, Mertely J, Legard DJPD. Comparison of epidemics of botrytis fruit rot and powdery mildew of strawberry in large plastic tunnel and field production systems. Plant Disease. 2001;**85**(8):901-909

[34] Wedge DE, Smith BJ, Quebedeaux JP, Constantin RJJCP. Fungicide management strategies for control of strawberry fruit rot diseases in Louisiana and Mississippi. Crop Protection. 1 Sep 2007;**26**(9):1449-1458

[35] Vellicce GR, Ricci JCD, Hernández L, APJT C. Enhanced resistance to Botrytis cinerea mediated by the transgenic expression of the chitinase gene ch5B in strawberry. Transgenic Research. Feb 2006;**15**:57-68

[36] Fernández-Ortuño D, Grabke A, Li X, Schnabel GJP. Independent emergence of resistance to seven chemical classes of fungicides in Botrytis cinerea. Phytopathology. 8 Apr 2015;**105**(4):424-432

[37] Rupp S, Weber RW, Rieger D, Detzel P, MJF H. Spread of Botrytis cinerea strains with multiple fungicide resistance in German horticulture. Frontiers in Microbiology. 3 Jan 2017;7:2075

[38] Alengebawy A, Abdelkhalek ST, Qureshi SR, MQJT W. Heavy metals and pesticides toxicity in agricultural soil and plants: Ecological risks and human health implications. Toxics. 25 Feb 2021;**9**(3):42

[39] Huang R, Li G, Zhang J, Yang L, Che H, Jiang D, et al. Control of postharvest botrytis fruit rot of strawberry by volatile organic compounds of Candida intermedia. Phytopathology. Jul 2011;**101**(7):859-869

[40] Sylla J, Alsanius BW, Krüger E, WJE W. Control of Botrytis cinerea in strawberries by biological control agents applied as single or combined treatments. European Journal of Plant Pathology. Nov 2015;**143**:461-471

[41] Xu X-M, Jeger MJP. Theoretical modeling suggests that synergy may result from combined use of two biocontrol agents for controlling foliar pathogens under spatial heterogeneous conditions. Phytopathology. Aug 2013;**103**(8):768-775

[42] Schestibratov K, Dolgov SJSH. Transgenic strawberry plants expressing a thaumatin II gene demonstrate enhanced resistance to Botrytis cinerea. Scientia Horticulturae. 1 Sep 2005;**106**(2):177-189

[43] Vellicce GR, Ricci JC, Hernández L, Castagnaro AP. Enhanced resistance to Botrytis cinerea mediated by the transgenic expression of the chitinase gene ch5B in strawberry. Transgenic Research. 2006;**15**:57-68

[44] Kumari C, Sharma M, Kumar V, Sharma R, Kumar V, Sharma P, et al. Genome editing technology for genetic amelioration of fruits and vegetables for alleviating post-harvest loss. Bioengineering. 2022;**9**(4):176

[45] López-Casado G, Sánchez-Raya C, Ric-Varas PD, Paniagua C, Blanco-

Portales R, Muñoz-Blanco J, et al. CRISPR/Cas9 editing of the polygalacturonase FaPG1 gene improves strawberry fruit firmness. Horticulture Research. Mar 2023;**10**(3):uhad011

[46] García-Robles JM, Medina-Rodríguez LJ, Mercado-Ruiz JN, RJRIdTP B-S. Evaluación de desinfectantes para el control de microorganismos en frutas y verduras. Revista Iberoamericana de Tecnología Postcosecha. 2017;**18**(1):9-22

[47] Almeida MLB, Moura CFH, Innecco R, Santos AD, de Miranda FR. Postharvest shelf-life and fruit quality of strawberry grown in different cropping systems. Academic Journals. 2015;**10**:4053-4061

[48] Avis T, Martinez C, RJC T. Effect of chlorine atmospheres on the development of rhizopus rot [Rhizopus stolonifer] and gray mold [Botrytis cinerea] on stored strawberry fruits. Canadian Journal of Plant Pathology. 1 Dec 2006;**28**(4):526-532

[49] Tomas-Grau RH, Requena-Serra FJ, Hael-Conrad V, Martínez-Zamora MG, Guerrero-Molina MF, JCJP D-R. Soft mechanical stimulation induces a defense response against Botrytis cinerea in strawberry. Plant Cell Reports. Feb 2018;**37**:239-250

Chapter 5

Native *Trichoderma* Strains Biocontrol Potential against Soil-Borne Pathogens: Strawberry

Yunus Korkom

Abstract

Strawberry production remains important in the world. Soil-borne fungal pathogens (such as *Macrophomina phaseolina*, *Rhizoctonia* spp., *Fusarium oxysporum*, *Phytophthora* spp., and *Pythium* spp.) are causing serious problems for strawberry farmers. Distinct treatments, such as fumigation, resilient varieties, solarization, rotating crops, synthetic fungicides, and cultural practices are used to combat infections of soil-borne in strawberries. Since strawberry fruits are consumed immediately, fungicide treatments raise a number of problems, including pesticide residue on the fruits which gives harmful effects on consumers. Solarized soils are often effective against certain soil-borne pathogens. New studies have focused on eco-friendly biological control agents (BCAs) that can be used as effective substitutes for fungicides. *Trichoderma* strains are efficient BCAs that have different mechanisms against soil-borne diseases in strawberries. Despite the success of commercial *Trichoderma*-based products, their low efficacy or ineffectiveness against targeted pathogens are major limitations under field conditions. Native *Trichoderma* strains that can be used to control this disease are ideal antagonists. This section discusses the potential of native *Trichoderma* strains to combat soil-borne pathogens in strawberry fields.

Keywords: *Macrophomina*, *Rhizoctonia*, *Phytophthora*, *Fragaria × ananassa*, biological control

1. Introduction

Strawberry (*Fragaria × ananassa* Duchesne ex Rozier) have been around since ancient and is a member of Rosaceae family [1]. China is one of the top manufacturers of strawberries, in addition to the US, Türkiye, Mexico, Egypt, and Spain are significant producers. In 2023, there were 7.9 billion people on the planet, and by 2050, it's predicted that there would be about ten billion [2]. The need for more readily accessible food supplies is driven by the increase in population intensity. Abiotic factors can significantly harm agricultural crop systems. For example, worldwide warming and climate change are significant factors in these factors [3]. Fungi, bacteria, viruses, and other parasites are caused an important infection in crop production [4]. Although soil-borne plant diseases are extensively dispersed, only a few species exhibit a pattern of limited distribution. At the same time, crops may be at risk if pests, which serve as vectors for spreading plant infections, change their distribution patterns [5]. The prevalent strawberry pathogens enclosing *Macrophomina phaseolina*, *Rhizoctonia solani*, *Fusarium* spp., *Phytophthora fragariae*, *P. cactorum*, *Pythium* spp., *Verticillium dahliae, Botrytis cinerea*, *Colletotrichum acutatum*, *Colletotrichum gloeosporioides*, *Sphaerotheca macularis*, *Phomopsis obscurans*, *Gnomonia comari*, and *Xanthomonas fragariae* [6–11]. These pathogens are frequently difficult to control. They are free-living organisms with a diverse variety of hosts and the ability to thrive for extended periods on soil organic materials and plant detritus. However, these pathogens of fungi have resistant structures that can be produced even in the absence of plants, such as sclerotia, microsclerotia, oospores, or chlamydospores. Additionally, owing to the likeness of symptoms, diagnosis is challenging and time-consuming [12]. Strawberry growers use synthetic fumigants and chemical fungicides in an orderly round the production season to reduce soil-borne diseases. The extensive use of fumigants and fungicides has a negative impact on soil and warm-blooded [13]. Cultural precautions, such as crop rotation, bio-fumigation, anaerobic soil disinfestation, and soil solarization, are embraced by farmers; however, these methods have inconsistent results and are ineffective in soil-borne disease management. In addition, the increase in crop diversity in agriculture has necessitated the creation of new research strategies [14]. Biological control can be an alternative for the management of plant diseases. The *Trichoderma* genus (Hypocreales, Ascomycota) is extensively used to control soil-borne diseases and to promote growth in different plants [10, 15, 16]. They have several antagonistic tools such as mycoparasitism [17], antibiosis [18], production of cell wall-degrading enzymes [19], rhizosphere and root colonization [20], and nutrient competition [17]. *Trichoderma* is widely used in bio-remediation in pollutants of soils and waters [17], as regional and systemic resistance-inducing in plants [21]. *Trichoderma* species or isolates can show different levels of effectiveness in different geographic locations, diverse crops, or cultivar functions [22–24]. Therefore, it is important to use native *Trichoderma* isolates for protection from soil-borne diseases in the field. This review informs the use of native *Trichoderma*, with some perspectives on the biocontrol potential of strawberry production.

2. Strawberry production and soil-borne diseases

2.1 Cultural practices of strawberry production

Strawberries are cultivated commercially in a variety of settings, including open fields, polytunnels, and glasshouses [25]. High tunnel production is employed to extend the strawberry growing season and yield earlier than open-field production. Greenhouses have been enabled for a year in strawberry production [26]. The interval on strawberry plants is usually 30 cm by 30 cm and 1.5 m between each bed in the field [27]. Many different factors are considered when selecting a variety of strawberries. These include locations, production methods, and customer choices. Strawberry plants can be divided into three primary groupings based on their flowering behaviors. Plants of long-day begin flowering under long-day conditions and generate twice harvests each season. June-bearing strawberries, commonly referred to as short-day strawberries, give a start of flowering under short-day conditions. These plants are provided only one harvest per season. Day-neutral cultivars depend on temperature differences for flower commence and day-long is not important. Strawberry production in the field is preferred of day-neutral cultivars [28]. According to the growth stage, strawberry plants require between −0.016 and 0.032 L of water per week [29]. When growing strawberries, it is crucial to know when and how much irrigation is used so that the fruit receives the necessary moisture while preserving labor and water resources [30–32].

2.2 Soil-borne diseases in strawberry

Serious soil-borne diseases influencing strawberry production are Phytophthora root rot (*Phytophthora* spp.), black root rot (*Pythium* spp., *Rhizoctonia solani*), Fusarium wilt (*Fusarium oxysporum*), charcoal rot (*Macrophomina phaseolina*), and Verticillium wilt (*Verticillium dahliae*). These pathogens have spread all over the world [7, 33–35]. Without disease management, these soil-borne diseases led to 20–30% output reductions in strawberry yield [36, 37]. The use of an overhead irrigation method causes early-season outbreaks of Oomycete-induced diseases in annual strawberry production [38, 39]. Different species of *Phytophthora* (Stramenopila, Peronosporaceae) cause crown rot in strawberries. These species as *P. cactorum, P. fragariaefolia, P. citricola, P. nicotianae,* and *P. citrophthora* [40–42]. In the studies carried out, *Phytophthora* was initially determined only from soil or crown could. But as a result of the studies carried out in subsequent processes, it was determined that isolates obtained from other host plants and strawberry fruit also caused leather rot [43, 44]. *Phytophthora cactorum* was first recorded as ground leather rot on fruits and crown rot in strawberries [44]. Owing to the continual presence of strawberry plants in the cultivation of perennials and plantations, the source of inoculum might include oospores [45]. The first symptoms appear at the onset of bluish-green color on new leaves, after which the plant begins to wilt. Plants wilt and die within days due to disease progression and crown rot [46]. *Phytophthora fragariae* causes red core, which can be leading to plant death in strawberry-cultivated regions [47]. The definition of the disease indicates dark reddish-brown staining of crowns that begins at the upper or lower and the breakdown of vascular texture. Since secondary infections show rare throughout the one crop cycle, the disease develops in a monocyclic pattern. The parent inoculum source was an infected plantation. Additionally, soil fumigation prior to planting is sometimes unsuccessful in eradicating this disease in fruit-producing areas [39, 48]. Most strawberry varieties are not tolerant to *Phytophthora*. Black root rot is a common disease that causes the death of feeder roots and degrades the structural roots in strawberries. It limits productivity and affects strawberry production worldwide [49, 50]. The genera *Rhizoctonia* and *Pythium* largely cause the black root rot complexes [51, 52]. *Rhizoctonia* spp. are associated with the black root rot complex and fall into an imperfect Basidiomycete fungus [53]. *Rhizoctonia* isolates are classified as multinucleate or mononucleate [54]. *R. solani* is the high virulent than *R. fragariae* in strawberry plants [55, 56]. Soil fumigation is still the only effective treatment for root rot by *Rhizoctonia* in strawberry [57]. The early symptoms exhibit only a few plants, which are mostly found in poorly drained and compacted soil. Plants that have been affected lack feeder roots, and many of the bigger roots have broken off at rotten locations where the cortical tissue has collapsed. Because of root rot, some plants lose their lives throughout the growing season, whereas those that survive will be stunted and will yield fewer strawberries [58]. The genus *Pythium* (Pythiaceae, Oomycota) has 327 described species [59]. Some species were only recently identified from *Phytopythium* in strawberries [38, 60]. Fusarium wilt is an important disease of strawberries worldwide. This disease was reported *Fusarium oxysporum* f. sp. *fragariae* in several countries [34, 61–63]. The first symptom of

Fusarium is common to flower or fruit set in well-grown of strawberries. Here, the oldest leaves starting wilt, become gray-green, and then dry out [64]. *Macrophomina phaseolina* (Botryosphaeriaceae), a soil-borne phyopathogen fungus, is found worldwide and affects over 500 plant species in 100 families. Diseases including charcoal rot, rot in the stem as well as root, and seedling blight are brought on by this pathogen in strawberry [65]. The main infectious source of *M. phaseolina* is microsclerotia. This resistant construction can last in the soil for 15 years [66]. The intensity of disease increases when temperatures of aerial and soil reach 30–35°C and lowly 60% of soil humidity [67]. The germinating hyphae of *M. phaseolina* are able to infect roots during the seedling phase. Afterward, the fungus influences the vascular system and distorts the transfer of water and nutrition in the plant. Yield loss is related amount of inoculum in the soil at the same time intensity with the severity of the disease [67]. *Verticillium* genus in the phylum Ascomycota accounts for the vascular wilt of many plants. The species *V. dahliae* and *V. albo-atrum* cause serious economic losses worldwide [68]. *Verticillium dahliae* is common in strawberry-growing areas, and it can be particularly severe when springtime temperature fluctuations might stress the plants [69]. The conidia, or microsclerotia, of the pathogen, germinate in the being of root secretions; upon this, the germ tubes penetrate the plant roots. The pathogen can survive for more than ten years in the soil or as microsclerotia (ms) on tissues of dead plants [70–72]. Microsclerotia concentration is 2 ms/g soil that can rise 100% wilt in strawberries [73]. Therefore, this pathogen is rough to struggle with present management strategies. Some fungicides are utilized based on the targeted pathogens, including azoxystrobin, carbendazim, prochloraz, metalaxyl, mancozeb, difenoconazole, pyraclostrobine, dimethomorph, chlorothalonil, tebuconazole, and fludioxonil [74]. Azoxystrobin, metalaxyl, and difenoconazole due to the often use of fungicides, these pathogens have developed resistance to these fungicides. As a result, chemical combat is unsuccessful in offering enough control for the disease [75, 76]. There is no one-size-matches-all-solution to disease control in soil-based strawberry growing, and a combination of strategies can be useful. The soil fumigant Methyl Bromide (MB) controls a broad wide of pathogens, pests, and weeds, so farmers have preferred it often. MB was phased out in 2005 below the Montreal Protocol because of its impacts as the chief ozone stratum thinner, notwithstanding its effectiveness [77]. Holmes et al. [78]. summarized MB options with the inclusion of recent fungicides, stubborn varieties, solarization, anaerobic soil disinfection, plastic films, and cover crops. Soil fumigants used to substitute MB comprise 1–3-dichloropropene (1,3-D), Chloropicrin (Pic), Dazomet (DZ), and Metam Sodium (MS), Metam Potassium (MP) [26]. Solarization is used as an alternative treatment for pathogens in tropical regions. Solarization involves wetting the soil and then covering it with a transparent film to raise the soil temperature during the summer. The transparent film remains closed for 4–6 weeks. Yet solarization is not practicable in loud-altitude regions where strawberry production occurs [79].

3. Biocontrol mechanisms of *Trichoderma*

Trichoderma species have different mechanisms, which are briefly explained as follows. The mycoparasitism mechanism of *Trichoderma* takes place with many sequential actions: (i) positive chemotrophic growth toward the host, (ii) direct contact with the phytopathogenic fungus, and (iii) coiling around the host hyphae. *Trichoderma* also undergoes morphological changes in hyphae, which enables it to penetrate the host hyphae and death its biomass. These changes include the formation of appressorium-like penetration structures and the production of cell walls that degrade hydrolytic enzymes (CWDEs) [19]. CWDEs include chitinase, glucanase, N-acetylglucosaminidase, and protease [80]. The second biocontrol mechanism of *Trichoderma* is antibiosis. *Trichoderma* species generate a wide range of small-molecule compounds as secondary metabolites (SMs), including pyrones, terpenoids, steroids, and polyketides. Additionally, *Trichoderma* produces siderophores and a significant amount of peptaibiotics known as peptaiboles, which frequently include non-standard amino acids [18]. The 6-pentyl-alpha-pyrone (6-PP) is one of the most studied in Pyrones group. This compound has been associated with yellow pigmentation and a coconut aroma for production in some strains of *Trichoderma*. Moreover, this compound exhibited antimicrobial activity. Polyketides (PKs) have antimicrobial activity. The role of PKs has to simplify competition for nutrients, reduce the ability of pathogens, and set chemical communication with organisms [81]. Another biocontrol mechanism is competition. *Trichoderma* species' ability to displace other fungal species in the rhizosphere is that they are excellent contestants for ground and nutrients. *Trichoderma* strains that efficiently absorb nutrients and develop more quickly than their rivals will have a clear advantage in their ability to colonize and thrive in various habitats [82]. Another mechanism involves an increase in plant growth. *Trichoderma* species realize root colonization, plant nutrition, and growth, as well as enhance plant resistance to abiotic stresses [83]. *Trichoderma* has been observed to induce systemic resistance for both monocots and dicots plants. This reaction includes plant recognition of the fungus through systemic induced resistance (ISR), which is mediated by the phytohormones ethylene (ET) and jasmonic acid (JA) [17]. *Trichoderma* also stimulates the expression of genes related to pathogenesis (PR), which is mediated by salicylic acid (SA). This reaction, also brought on by biotrophic and hemibiotrophic infections, is referred to as systemic acquired resistance (SAR) [84]. The pollution of water and soil is occurring taking by extreme pesticides and synthetic fertilizer utilization in plant production fields. Chemical substances can be removed using either methods of biological or chemical [85]. In recent research, *Trichoderma* species have executed their skills to remediation on different fungicides, insecticides, and herbicides [86, 87]. This biological control agent successfully detoxifies into sulfonylurea (herbicide) [86], dichlorvos (insecticide) [88], carbendazim [89], and penthiopyrad [90].

4. The use of the antagonistic properties of native *Trichoderma* species against soil-borne pathogens in strawberry

In Giza, Egypt, a study using four distinct *Trichoderma* species to treat strawberry black root rot was carried out. The pathogen isolates *F. solani*, *R. solani*, and *Pythium* sp. employed in the investigation was isolated from strawberry growing areas. Black root rot was assessed in vitro and in field settings using native *T. harzianum*, *T. viride*, *T. virinis*, and *T. koningii* isolates. The *in vitro* investigation revealed that the pathogen isolates had their mycelial development inhibited by *T. harzianum* 92.1, 91.7, 93.4%, *T. viride* 91.7, 92.9, 90.3%, *T. virinis* 87.4, 86.6, 84.3%, *T. koningii* 92.1, 93.4, 90.6%, respectively. A variety of Festival strawberries was evaluated for plant growth, yield, and disease severity in field conditions. The combination of all *Trichoderma* isolates produced the greatest rise in the fresh and dry weight by 83.3 and 176.9%, respectively. Moreover, the mix of *Trichoderma* species showed the highest increase in yield (117.1%). Single-proceedings of *T. harzianum*, *T. viride*, and *T. koningii* raised the fresh weight of plants by 105 to 68%, and these isolates showed increasing yield by 71.1, 57.1, and 64.3% respectively. *T. virinis* was less efficient in fresh-dry weight and yield. The Trichoderma mix treatment enhanced the peroxidase and chitinase activity in strawberry leaves by 150 and 160.9%, respectively [91]. *Rhizoctonia solani, Fusarium solani, F. oxysporum, and Macrophomina phaseolina of* diseased strawberry plants were collected from Qalyubia Governorate, Egypt. *Trichoderma album*, *T. harzianum*, *T. viride*, and *T. hamatum* were isolated from rhizospheric soils in the roots of healthy strawberry plants. *In vitro* test results, all pathogens were reduced growth by *T. harzianum* 83.43, 78.53, 67.10, 69.50%, *T. album* (77.53, 72.67, 66.53, 68.43%), *T. hamatum* (55.70, 63.27, 56.53, 53.33%), *T. viride* (74.57, 70.03, 65.47, 61.23, 67.83%), respectively. Field treatment results of disease incidence (7-weeks post transplanting), *T. harzianum* showed the highest impact (77.0, 67.2%) follow up by *T. album* (72.0, 63.4%), *T. hamatum* (27.0, 29.9%) in the 2015/16 and 2016/17 seasons, respectively. *T. harzianum* was the most successful treatment in terms of chlorophyll, protein, total nitrogen, total phenols, and total sugars. Moreover, this treatment greatest increased in yield (36.84% and 25.22%) when compared to the control treatment over the two consecutive growing seasons [92]. *Trichoderma harzianum* strain T-H4 inhibited mycelial growth on *Rhizoctonia* sp. 70.22 ± 5.46, and *Fusarium* sp. 63.65 ± 1.50 in dual culture test [93]. Mirmajlessi et al. [94] isolated *V. dahliae* (29 isolates) from strawberry-growing areas in Estonia. *V. dahliae* isolate SV-19 had high virulence (79.51%). Volatile metabolites of *T. harzianum* isolates showed different inhibitory values on *V. dahliae*, from 15.8% to 88.3%, while TU79 determined the maximum effect on the pathogens. This isolate had significant mycelial inhibition in dual culture (92.1%) and non-volatile metabolites. The results of *in vitro* studies chosen seven isolates (TU63, TU68, TU72, TU74, TU75, TU79, and TU80) (mycelial inhibition rate > 50%) for the greenhouse treatment. In greenhouse studies, the antagonist isolates were observed with different rates of disease severity from 14.2% to 75.6%. The isolate of TU79 represented only the highest reduced disease severity (≈27%) that was applied to the soil or root [94]. The efficiency of four *T. harzianum* was analyzed against *F. oxysporum* f. sp. *fragariae* using a dual culture approach. Solely one strain of *T. harzianum* observed to inhibit pathogen growth. This isolate treatment protected strawberry plants (80.0%) in the greenhouse after 42 days [95]. The present study identified harzianic acid produced by *T. harzianum* M10, and hydrophobin produced by *T. longibrachiatum* MK1 effect on strawberry plant (cv. Sabrina) growth in Naples, Italy. The plant-growing season started in October 2016 and ended in June 2017, and the fruits were harvested once each week from April to June 2017. The harzianic acid increased by 24% in total fruit yield (g plant^{-1}) and it had the maximum number of fruits for each plant (14%). Also, this treatment showed the highest increase in the total soluble solids content (8%). The hydrophobin had effectiveness on the ascorbic acid content (9%), and it showed an increment in root length (15%), root fresh weight (15%), and dry weight (19%). The maximum content of cya 3-O-glc and pel 3-O-rut was provided by hydrophobin as 63%, 11%, respectively [96]. Native *T. citrinoviride* determined the ability to biocontrol of *R. solani* (isolated from strawberry root) in dual culture and pot trial by Sekmen Cetinel et al. [97]. The result of the dual culture test, *T. citrinoviride* showed a high inhibition of 79% against *R. solani*. *T. atroviride* + *R. solani* using determined disease incidence from 33% to 41%, and *Trichoderma* pre-treatment increased the dry weight of strawberry plant (cv. Rubigen) in pot trial. Moreover, this treatment indicated a higher plane of PSII. *Trichoderma* isolates, *R. fragariae,* and *R. solani* were isolated from different strawberry fields regions in Menoufia, Egypt. *T. harzianum* and *T. hamatum in vitro* test showed different inhibition values on *R. fragaria* by 83.3%, 72.2%, respectively. This biocontrol agent effect in field trials determined on disease severity of *R. fragaria of* strawberry plants (cv. Sana) by 9.6%, 6.8%, respectively. The group control had a disease severity of 86.4%. *T. harzianum* and *T. hamatum* increased yield (g/plant) by 50.4%, 33.2%, respectively [98]. *R. solani* and native *Trichoderma* isolates isolated from strawberry roots and soils in Qingdao, China. These *Trichoderma* isolates are namely *T. atroviride* T1, *T. harzianum* T2, *T. atroviride* T3, *T. harzianum* T4, *T. harzianum* T5, respectively. In the dual, non-VOC, and VOC assays, T1 and T3 had maximum inhibition rates of 60.4–100–75.3%, 60.9–100–69.8%, respectively. All native *Trichoderma* isolates showed well-enzymatic activities of protease, cellulose, chitinase, and glucanase. Another result obtained from this study is that the antagonistic effect of the five *Trichoderma* isolates against *R. solani* in the petioles of strawberries was ordered as follows: T2 (34.8%) > T4 (−16.7%) > T3 (−46%) > T5 (−171.5%) > T1 (−172%) [99]. Native *T. viride* isolates were collected from the rhizosphere of sound strawberries. The pathogen isolate was *R. fragariae* which was isolated in the northwest of Egypt in Ismailia district. The antagonist inhibited the mycelial growth of *Rhizoctonia* by 69.5%. The ethylene synthesis increased when the assessment 2–7 days after the process in plant assays of *R. fragariae* + *T. viride* treatment. The maximum SOD activity was considered at day 6 (T3, T4, and T5), increased at day 7, and stayed high at day 8 on the treatment of T3, T4, and T5 in field and greenhouse studies. The T3 treatment showed superlative CAT activity on day 4, while T4 and T5 remained stable until the end of the field and greenhouse trials. The *Trichoderma* treatments of plant fruit numbers showed a difference: T3, T4, and T5 had maximum fruit but T2 had minimum fruit in the greenhouse and field. In strawberry plants treated with *Trichoderma*, there was a lower disease incidence and a bulkier root than in the other plants [100]. Different isolates of *T. viride*, *T. viride*, *T. hamatum*, *T. koningii*, and *T. harzianum* species were obtained from soil samples taken from strawberry production areas in Buhayre province in Egypt. These isolates were evaluated against *M. phaseolina* (M1 and M3 isolates) isolated from strawberries showing symptoms of charcoal rot disease from the same region under *in vitro* and greenhouse conditions. As a result of the dual culture test, *T. viride* isolates had the highest inhibition rate against *M. phaseolina* isolates. As a result of the study conducted under greenhouse conditions, it was reported that *T. hamatum* (1, 2) application had the best effect on the survival rate of pathogen-infected plants [101]. Korkom and Yildiz [102] investigated ten *Trichoderma* isolates isolated from the soil of strawberry production areas in Aydin province, Türkiye, and their obtained effects on *M. phaseolina* under in vitro and in vivo conditions. Within the scope of antagonistic studies carried out *in vitro*, it was determined that *Trichoderma* isolates limited the mycelial growth of *M. phaseolina* by 25.9–59.1% in dual culture and VOCs formed by *Trichoderma* isolates by 13.3–30.0%. In addition, it was determined that *Trichoderma* isolates reduced microsclerot formation in the dual culture assay by 11.7–63.1% and the effect of volatile compounds by 9.7–77.3%. In vivo, the study was carried out by simultaneous application of antagonists and pathogen to strawberry plants (cv. Rubygem) (Tr + Mp) and pathogen inoculation 15 days after antagonists application (Tr + Mp15). Tr28 showed the highest weight rate (36.47%) and no plant death was observed in Tr + Mp. In Tr + Mp(15), Tr25 was observed the highest wet weight rate (47.37%) and plant death did not occur in Tr26, Tr24, Tr21, Tr28. *T. harzianum* (sixteen isolates), *T. virens* (three isolates) were obtained from the soil in strawberry production areas of Aydin province, Türkiye. The effects of these isolates on *M. phaseolina* and plant growth in strawberries (cv. Festival) under *in vitro* and *in vivo* conditions. Tvr4 (*T. virens*) inhibited the highest rate (66.4%) of mycelial growth of the pathogen in the dual culture test. The VOCs produced by Tvr2 limited the mycelial growth of the pathogen (33.3%). Thr15 (*T. harzianum*) and Tvr4 reduced the highest rate of microsclerot formation by 80.5%, 73%, respectively. Thr15 provided the highest plant wet weight rate (281.95%) when the antagonists and pathogen were applied to the soil at the same time in vivo assay. Furthermore, no mortality occurred in plants treated with Thr13 and Thr14 isolates. Thr20 was determined to be the isolate with the highest plant wet weight rate (411.47%) in the pathogen when it was applied to the soil 15 days after the antagonist's application. No mortality occurred in the plants applied at Thr13 and Thr14. Thr22 and Thr6 provided high wet weight rates of 344.93%, 331.15%, respectively, in the plant growth assay [103]. The isolation and identification of *Trichoderma* spp. from different agricultural areas of Aydin province, Türkiye. Eighty-eight *Tichoderma* isolates were obtained from 165 soil samples in Aydin province. *T. afroharzianum*, *T. guizhouense*, and *T. harzianum* by morphological and molecular identification in this study of *Trichoderma* isolates. The *T. guizhouense* isolates obtained in this study are the first records for Türkiye. These antagonistic isolates were carried out to determine their biological control potential against charcoal rot disease and plant growth in strawberries *in vitro*, *in vivo*, and in field conditions. *In vivo* and field studies were carried out on the Fortuna strawberry variety, and the trials were repeated for two years. As a result of the study conducted under greenhouse and field conditions, the TrMix application had the best effect on plant mortality (mean 14.1%) and yield (mean 29.7%). In field studies, native *T. afroharzianum* isolates were applied five times during strawberry production per season and were successfully colonized in the soil. Moreover, all *Trichoderma* isolates showed the highest enzymatic activities, including chitinase, sellulase, and sidephore [35].

5. Conclusions

From the perspective of agriculture, the main issues surrounding soil and environmental health include an increase in pesticide residue in soil and water, a decrease in soil-beneficial microorganisms, changes to the physical and chemical makeup of the soil, and pesticide resistance of the pathogen. This chapter of the book compiled how novel *Trichoderma* species control soil-borne diseases in strawberries. Many studies have indicated that *Trichoderma* species have multiple beneficial effects on strawberries, that not only include disease control, but also the stimulation of plant growth, increased yield, uptake of nutrients, and healing of crop quality. Therefore, the interactions that take place between soil-borne pathogens, micro- or macroorganisms, and even the physicochemical environmental conditions, it is crucial to thoroughly comprehend these interactions as possible for disease management measures to apply them as effectively in strawberry production. Additionally, *further research* is needed to develop models for different systems to assess the application of native *Trichoderma* isolates in strawberries.

Author details

Yunus Korkom
Faculty of Agriculture, Department of Plant Protection, Aydın Adnan Menderes University, Aydin, Turkey

*Address all correspondence to: yunus.korkom@adu.edu.tr

References

[1] Padmanabhan P, Mizran A, Sullivan JA, Paliyath G. Strawberries. In: Caballero B, Finglas PM, Toldrá F, editors. Encyclopedia of Food and Health. USA: Elsevier; 2016. pp. 193-198. DOI: 10.1016/B978-0-12-384947-2.00667-X

[2] Food and Agriculture Organization of the United Nations (FAO). The State of Food and Agriculture. Strawberry Production [Internet]. 2021. Available from: http://www.fao.org/faostat/es/#data/QC/visualize [Accessed: June 28, 2023]

[3] Cooley S, Schoeman D, Bopp L, Boyd P, Donner S, Ito SI, et al. Oceans and coastal ecosystems and their services. In: IPCC AR6 WGII. England: Cambridge University Press; 2022. pp. 379-550

[4] Rodrigues AO, May De Mio LL, Soccol CR. *Trichoderma* as a powerful fungal disease control agent for a more sustainable and healthy agriculture: Recent studies and molecular insights. Planta. 2023;**257**(2):31. DOI: 10.1007/s00425-022-04053-4

[5] Shipley JR, Twining CW, Taff CC, Vitousek MN, Flack A, Winkler DW. Birds advancing lay dates with warming springs face greater risk of chick mortality. Proceedings of the National Academy of Sciences. 2020;**117**(41):25590-25594. DOI: 10.1073/pnas.200986411

[6] Avilés M, Castillo S, Bascon J, Zea-Bonilla T, Martín-Sánchez PM, Pérez-Jiménez RM. First report of *Macrophomina phaseolina* causing crown and root rot of strawberry in Spain. Plant Pathology. 2008;**57**(2):382. DOI: 10.1111/j.1365-3059.2007.01717.x

[7] Murthy BNS, Pramanick KK. Strawberry cultivation in mild-tropics: Prospects and challenges from diseases' perspective. Acta Horticulture. 2014;**1049**:151-159. DOI: 10.17660/ActaHortic.2014.1049.13

[8] Lafarga T, Colás-Medà P, Abadías M, Aguiló-Aguayo I, Bobo G, Viñas I. Strategies to reduce microbial risk and improve quality of fresh and processed strawberries: A review. Innovative Food Science & Emerging Technologies. 2019;**52**:197-212. DOI: 10.1016/j.ifset.2018.12.012

[9] Borrero C, Capote N, Gallardo MA, Avilés M. First report of vascular wilt caused by *Fusarium proliferatum* on strawberry in Spain. Plant Disease. 2019;**103**(3):581-581. DOI: 10.1094/PDIS-09-18-1544-PDN

[10] Oliveira MS, Wang NY, Peres NA. Multilocus phylogenetic analyses of *Colletotrichum gloeosporioides* species complex causing crown rot on strawberry in Florida. Phytopathology. 2022;**112**(4):898-906. DOI: 10.1094/PHYTO-04-20-0151-R

[11] Mazzola M, Muramoto J, Shennan C. Anaerobic disinfestation induced changes to the soil microbiome, disease incidence and strawberry fruit yields in California field trials. Applied Soil Ecology. 2018;**127**:74-86. DOI: 10.1016/j.apsoil.2018.03.009

[12] Patil L, Thomse S, Ghante PH, Pawar S. Sustainable management of soil-borne plant diseases. AgriCos e-newsletter. 2021;**2**(3):98-101

[13] Panth M, Hassler SC, Baysal-Gurel F. Methods for management of soilborne diseases in crop production. Agriculture.

2020;**10**(1):16. DOI: 10.3390/agriculture10010016

[14] Ons L, Bylemans D, Thevissen K, Cammue BP. Combining biocontrol agents with chemical fungicides for integrated plant fungal disease control. Microorganisms. 2020;**8**(12):1930. DOI: 10.3390/microorganisms8121930

[15] Woo SL, Ruocco M, Vinale F, Nigro M, Marra R, Lombardi N, et al. *Trichoderma*-based products and their widespread use in agriculture. The Open Mycology Journal. 2014;**8**:71-126. DOI: 10.2174/1874437001408010071

[16] Hewedy OA, Abdel Lateif KS, Seleiman MF, Shami A, Albarakaty FM, El-Meihy RM. Phylogenetic diversity of *Trichoderma* strains and their antagonistic potential against soil-borne pathogens under stress conditions. Biology (Basel). 2020;**9**(8):189. DOI: 10.3390/biology9080189

[17] Harman GE, Howell CR, Viterbo A, Chet I, Lorito M. *Trichoderma* species—opportunistic, avirulent plant symbionts. Nature Reviews Microbiology. 2004;**2**(1): 43-56. DOI: 10.1038/nrmicro797

[18] Degenkolb T, Gräfenhan T, Berg A, Nirenberg HI, Gams W, Brückner H. Peptaibiomics: Screening for polypeptide antibiotics (peptaibiotics) from plant-protective *Trichoderma* species. Chemistry & Biodiversity. 2006;**3**(6):593-610. DOI: 10.1002/cbdv.200690063

[19] Herrera-Estrella A, Chet I. The biological control agent *Trichoderma* from fundamentals to applications. Mycology Series. 2004;**21**:147-156

[20] Benítez T, Rincón AM, Limón MC, Codon AC. Biocontrol mechanisms of *Trichoderma* strains. International Microbiology. 2004;**7**(4):249-260

[21] Bigirimana J, De Meyer G, Poppe J, Elad Y, Höfte M. Induction of systemic resistance on bean (*Phaseolus vulgaris*) by *Trichoderma harzianum*. Mededelingen van de Faculteit Landbouwkundige en Toegepaste Biologische Wetenschappen, Universiteit Gent. 1997;**62**:1001-1007

[22] Tucci M, Ruocco M, De Masi L, De Palma M, Lorito M. The beneficial effect of *Trichoderma* spp. on tomato is modulated by the plant genotype. Molecular Plant Pathology. 2011;**12**(4):341-354. DOI: 10.1111/j.1364-3703.2010.00674.x

[23] Hautsalo J, Vestberg M, Parikka P, Kukkonen S, Karhu S, Tahvonen R. Biological control of strawberry crown rot is substrate dependent phenomenon. Journal of Berry Research. 2016;**6**(1):65-79. DOI: 10.3233/JBR-150110

[24] Bazghaleh N, Prashar P, Woo S, Vanderberg A. Effects of lentil genotype on the colonization of beneficial *Trichoderma* species and biocontrol of Aphanomyces root rot. Microorganisms. 2020;**8**:1290. DOI: 10.3390/microorganisms8091290

[25] Defterli SG. Review of robotic technology for strawberry production. Applied Engineering in Agriculture. 2016;**32**(3):301-318. DOI: 10.13031/aea.32.11318

[26] Hernández-Martínez NR, Blanchard C, Wells D, Salazar-Gutiérrez MR. Current state and future perspectives of commercial strawberry production: A review. Scientia Horticulturae. 2023;**312**:111893. DOI: 10.1016/j.scienta.2023.111893

[27] Cvelbar Weber N, Koron D, Jakopič J, Veberič R, Hudina M, Baša ČH. Influence of nitrogen, calcium and nano-fertilizer on strawberry (*Fragaria*×*ananassa* Duch.) fruit inner and outer quality.

Agronomy. 2021;**11**(5):997. DOI: 10.3390/agronomy11050997

[28] López-Aranda JM, Soria C, Santos BM, Miranda L, Domínguez P, Medina-Mínguez JJ. Strawberry production in mild climates of the world: A review of current cultivar use. International Journal of Fruit Science. 2011;**11**(3):232-244. DOI: 10.1080/15538362.2011.608294

[29] Morillo JG, Martín M, Camacho E, Díaz JR, Montesinos P. Toward precision irrigation for intensive strawberry cultivation. Agricultural Water Management. 2015;**151**:43-51. DOI: 10.1016/j.agwat.2014.09.021

[30] Létourneau G, Caron J, Anderson L, Cormier J. Matric potential-based irrigation management of field-grown strawberry: Effects on yield and water use efficiency. Agricultural Water Management. 2015;**161**:102-113. DOI: 10.1016/j.agwat.2015.07.005

[31] Gendron L, L'etourneau G, Depardieu C, Anderson L, Sauvageau G, Levallois R, et al. Irrigation management based on soil matric potential improves water use efficiency of field-grown strawberries in California. Acta Horticulturae. International Society for Horticultural Science. 2017;**1156**:191-196. DOI: 10.17660/ActaHortic.2017.1156.29

[32] Ariza MT, Miranda L, G'omez-Mora JA, Medina JJ, Lozano D, Gavil'an P, et al. Yield and fruit quality of strawberry cultivars under different irrigation regimes. Agronomy. 2021;**11**(2):261. DOI: 10.3390/agronomy11020261

[33] Benlioglu S, Yildiz A, Döken T. Studies to determine the causal agents of soil-borne fungal diseases of strawberries in Aydin and to control them by soil disinfestation. Journal of Phytopathology. 2004;**152**(8-9):509-513. DOI: 10.1111/j.1439-0434.2004.00888.x

[34] Koike ST, Kirkpatrick SC, Gordon TR. Fusarium wilt of strawberry caused by *Fusarium oxysporum* in California. Plant Disease. 2009;**93**(10):1077-1077. DOI: 10.1094/PDIS-93-10-1077A

[35] Korkom Y. Aydın ili tarım alanlarından elde edilen *Trichoderma* spp.'nin tanılanması ile çilekte kömür çürüklüğü hastalığına ve bitki gelişimine etkilerinin belirlenmesi [thesis]. Aydin Adnan Menderes University; 2022

[36] Carpenter JE, Gianessi LP, Lynch L. The Economic Impact of the Scheduled US Phaseout of Methyl Bromide. Washington DC: National Center for Food and Agricultural Policy; 2000. pp. 70-137

[37] Subbarao KV, Kabir Z, Martin FN, Koike ST. Management of soilborne diseases in strawberry using vegetable rotations. Plant Disease. 2007;**91**(8):964-972. DOI: 10.1094/PDIS-91-8-0964

[38] Marin MV, Seijo T, Mertely J, Peres NA. First report of crown rot caused by *Phytopythium helicoides* on strawberry in the Americas. Plant Disease. 2019;**103**:2696. DOI: 10.1094/PDIS-03-19-0658-PDN

[39] Baggio JS, Cordova LG, Toledo BF, Noling JW, Peres NA. A reassessment of the fungicidal efficacy of 1, 3-dichloropropene, chloropicrin, and metam potassium against *Macrophomina phaseolina* in strawberry. Pest Management Science. 2022;**78**(8):3416-3423. DOI: 10.1002/ps.6982

[40] Meszka B, Michalecka M. Identification of *Phytophthora* spp. isolated from plants and soil samples

on strawberry plantations in Poland. Journal of Plant Diseases and Protection. 2016;**123**:29-36. DOI: 10.1007/s41348-016-0007-2

[41] Marin MV, Seijo T, Oliveira MS, Zuchelli E, Mertely J, Peres NA. First report of *Phytophthora nicotianae* causing crown rot of strawberry in the United States. Plant Disease. 2018;**102**(7):1463-1463. DOI: 10.1094/PDIS-08-17-1333-PDN

[42] Farr DF, Rossman AY. Fungal databases, U.S. national fungus collections, ARS, USDA [Internet]. 2022. Available from: https://nt.ars-grin.gov/fungaldatabases

[43] Hantula J, Lilja A, Nuorteva H, Parikka P, Werres S. Pathogenicity, morphology and genetic variation of Phytophthora cactorum from strawberry, apple, rhododendron, and silver birch. Mycological Research. 2000;**104**(9):1062-1068. DOI: 10.1017/S0953756200002999

[44] Sharma S, Marin MV, Lee MB, Baggio JS, Peres NA, Lee S. Genomic approaches for improving resistance to Phytophthora crown rot caused by P. cactorum in strawberry (*Fragaria×ananassa*). Frontiers in Agronomy. 2022;**4**:941111. DOI: 10.3389/fagro.2022.941111

[45] Srivastava J, Singh A, Sharma R. Diseases of apples and their management. In: Diseases of Fruits and Vegetable Crops: Recent Management Approaches. Vol. 1. Boca Raton, FL, USA: Fruit Crops, CRC Press; 2022

[46] Peres NA, Baggio JS. Phytophthora crown rot of strawberry: PP350, 10/2019. EDIS. 2019;**5**:3-3

[47] Hughes KJD, Inman AJ, Cooke DEL. Comparative testing of nested PCR-based methods with bait-plant tests for detecting *Phytophthora fragariae* var. *fragariae* in infected strawberry roots from fruit crops in the UK. EPPO Bulletin. 2000;**30**(3-4):533-538

[48] Browne GT, Bhat RG. Phytophthora crown and root rot. California Strawberry Comm. 2011;**9**:1-4

[49] Mouden N, Al Batnan A, Benkirane R, Ouazzani Touhami A, Douira A. Diversity and distribution of fungi from strawberry plants grown in Gharb-Loukkos (Morocco). International Journal of Recent Scientific Research. 2016;**7**(10):13630-13641

[50] Errifi A, Touhami AO, Mouden N, El Batnan AM, El Hazzat N, Selmaoui K, et al. Pathogenicity of *Rhizoctonia solani* towards distinct varieties of cultivated strawberry plants in Morocco. Plant Cell Biotechnology and Molecular Biology. 2019;**20**:1086-1099

[51] Abad ZG, Louws FJ, Fernandez GE, Ferguson LM. Predominance and pathogenicity of fungi and stramenopiles associated with black root rot of strawberries. Phytopathology. 2022;**92**(6):S1

[52] Ferguson LM, Louws FJ, Abad ZG, Fernandez GE, Poling EB, Brannen PM. Impacts of alternatives on strawberry yield and root colonization by fungal pathogens. In: Annual International Research Conference on Methyl Bromide Alternatives and Emissions Reductions. San Diego. 3-6 Nov 2003. pp. 1-56

[53] Cubeta MA, Vilgalys R. *Rhizoctonia*. In: Lederberg J, editor. Encyclopedia of Microbiology. San Diego, CA: Academic Press; 2000. pp. 109-116

[54] Sharon M, Freeman S, Kuninaga S, Sneh B. Genetic diversity, anastomosis groups and virulence of *Rhizoctonia*

spp. from strawberry. European Journal of Plant Pathology. 2007;**117**:247-265. DOI: 10.1007/s10658-006-9091-7

[55] Martin FN. *Rhizoctonia* spp. recovered from strawberry roots in central coastal California. Phytopathology. 2000;**90**(4):345-353. DOI: 10.1094/PHYTO.2000.90.4.345

[56] Botha A, Denman S, Lamprecht SC, Mazzola M, Crous PW. Characterization and pathogenicity of *Rhizoctonia* isolates associated with black root rot of strawberries in the Western Cape Province, South Africa. Australasian Plant Pathology. 2003;**32**:195-201. DOI: 10.1071/AP02067

[57] Fang X, Finnegan PM, Barbetti MJ. Wide variation in virulence and genetic diversity of binucleate *Rhizoctonia* isolates associated with root rot of strawberry in Western Australia. PLoS One. 2013;**8**(2):e55877. DOI: 10.1371/journal.pone.0055877

[58] Asad-Uz-Zaman M, Bhuiyan MR, Khan MAI, Bhuiyan MKA, Latif MA. Integrated options for the management of black root rot of strawberry caused by *Rhizoctonia solani* Kuhn. Comptes Rendus Biologies. 2015;**338**(2):112-120. DOI: 10.1016/j.crvi.2014.11.006

[59] Uzuhashi S, Kakishima M, Tojo M. Phylogeny of the genus *Pythium* and description of new genera. Mycoscience. 2010;**51**(5):337-365. DOI: 10.1007/S10267-010-0046-7

[60] Ishiguro Y, Otsubo K, Watanabe H, Suzuki M, Nakayama K, Fukuda T, et al. Root and crown rot of strawberry caused by *Pythium helicoides* and its distribution in strawberry production areas of Japan. Journal of General Plant Pathology. 2014;**80**:423-429. DOI: 10.1007/s10327-014-0520-8

[61] Arroyo FT, Llergo Y, Aguado A, Romero F. First report of Fusarium wilt caused by *Fusarium oxysporum* on strawberry in Spain. Plant Disease. 2009;**93**(3):323-323. DOI: 10.1094/PDIS-93-3-0323B

[62] Williamson M, Fernández-Ortuño D, Schnabel G. First report of Fusarium wilt of strawberry caused by *Fusarium oxysporum* in South Carolina. Plant Disease. 2012;**96**(6):911-911. DOI: 10.1094/PDIS-02-12-0164-PDN

[63] Stanković I, Ristić D, Vučurović A, Milojević K, Nikolić D, Krstić B, et al. First report of Fusarium wilt of strawberry caused by *Fusarium oxysporum* in Serbia. Plant Disease. 2014;**98**(10):1435-1435. DOI: 10.1094/PDIS-04-14-0346-PDN

[64] Koike ST, Gordon TR. Management of Fusarium wilt of strawberry. Crop Protection. 2015;**73**:67-72. DOI: 10.1016/j.cropro.2015.02.003

[65] Ghosh TANMAY, Biswas MK, Guin C, Roy P. A review on characterization, therapeutic approaches and pathogenesis of *Macrophomina phaseolina*. Plant Cell Biotechnology and Molecular Biology. 2018;**19**:72-84

[66] Gupta GK, Sharma SK, Ramteke R. Biology, epidemiology and management of the pathogenic fungus *Macrophomina phaseolina* (Tassi) Goid with special reference to charcoal rot of soybean (*Glycine max* (L.) Merrill). Journal of Phytopathology. 2012;**160**(4):167-180. DOI: 10.1111/j.1439-0434.2012.01884.x

[67] Marquez N, Giachero ML, Declerck S, Ducasse DA. *Macrophomina phaseolina*: General characteristics of pathogenicity and methods of control. Frontiers in Plant Science. 2021;**12**:634397. DOI: 10.3389/fpls.2021.634397

[68] Pegg GF, Brady BL. Hosts. In: Verticillium Wilts. Wallingford UK: CABI Publishing; 2002. pp. 293-340

[69] Maas JL. Compendium of Strawberry Diseases. St. Paul: APS Press; 1998

[70] Fradin EF, Thomma BPHJ. Physiology and molecular aspects of Verticillium wilt diseases caused by *V. dahliae* and *V. albo-atrum*. Molecular Plant Pathology. 2006;**7**:71-86. DOI: 10.1111/j.1364-3703.2006.00323.x

[71] Lord CA, Smith CA. Verticillium Wilt of Strawberry. Fact Sheet. Durham, New Hampshire, USA: The University of New Hampshire Cooperative extension; 1994

[72] Klosterman SJ, Atallah ZK, Vallad GE, Subbarao KV. Diversity, pathogenicity, and management of *Verticillium* species. Annual Review of Phytopathology. 2009;**47**:39-62. DOI: 10.1146/annurev-phyto-080508-081748

[73] Harris DC, Yang JR. The relationship between the amount of *Verticillium dahliae* in soil and the incidence of strawberry wilt as a basis for disease risk prediction. Plant Pathology. 1996;**45**(1):106-114. DOI: 10.1046/j.1365-3059.1996.d01-96.x

[74] Xia Q, Rufty T, Shi W. Soil microbial diversity and composition: Links to soil texture and associated properties. Soil Biology and Biochemistry. 2020;**149**:107953. DOI: 10.1016/j.soilbio.2020.107953

[75] Abd El-Hai KM, El-Metwally MA, El-Baz SM, Zeid AM. The use of antioxidants and microelements for controlling damping-off caused by *Rhizoctonia solani* and charcoal rot caused by *Macrophomina phasoliana* on sunflower. Plant Pathology Journal. 2009;**8**(3):79-89. DOI: 10.3923/ppj.2009.79.89

[76] Ren C, Liu K, Dou P, Li J, Wang K. The changes in soil microorganisms and soil chemical properties affect the heterogeneity and stability of soil aggregates before and after grassland conversion. Agriculture. 2022;**12**(2):307. DOI: 10.3390/agriculture12020307

[77] Saltzman ES, Nicewonger MR, Montzka SA, Yvon-Lewis SA. A post-phaseout retrospective reassessment of the global methyl bromide budget. Journal of Geophysical Research: Atmospheres. 2022;**127**(3):e2021JD035567. DOI: 10.1029/2021JD035567

[78] Holmes GJ, Mansouripour SM, Hewavitharana SS. Strawberries at the crossroads: Management of soilborne diseases in California without methyl bromide. Phytopathology. 2020;**110**(5):956-968. DOI: 10.1094/PHYTO-11-19-0406-IA

[79] Hartz TK, DeVay JE, Elmore CL. Solarization is an effective soil disinfestation technique for strawberry production. HortScience. 1993;**28**(2):104-106. DOI: 10.21273/HORTSCI.28.2.104

[80] Tseng SC, Liu SY, Yang HH, Lo CT, Peng KC. Proteomic study of biocontrol mechanisms of *Trichoderma harzianum* ETS 323 in response to *Rhizoctonia solani*. Journal of Agricultural and Food Chemistry. 2008;**56**(16):6914-6922. DOI: 10.1021/jf703626j

[81] Mukherjee PK, Horwitz BA, Kenerley CM. Secondary metabolism in *Trichoderma*–a genomic perspective. Microbiology. 2012;**158**(1):35-45. DOI: 10.1099/mic.0.053629-0

[82] Monfil VO, Casas-Flores S. Molecular mechanisms of biocontrol in *Trichoderma* spp. and their applications in agriculture. In: Biotechnology and Biology of *Trichoderma*. USA: Elsevier; 2014. pp. 429-453. DOI: 10.1016/B978-0-444-59576-8.00032-1

[83] Windham MT. A mechanism for increased plant growth induced by *Trichoderma* spp. Phytopathology. 1986;**76**:518-521

[84] Shoresh M, Harman GE, Mastouri F. Induced systemic resistance and plant responses to fungal biocontrol agents. Annual Review of Phytopathology. 2010;**48**:21-43. DOI: 10.1146/annurev-phyto-073009-114450

[85] Zin NA, Badaluddin NA. Biological functions of *Trichoderma* spp. for agriculture applications. Annals of Agricultural Sciences. 2020;**65**(2):168-178

[86] Vázquez MB, Barrera V, Bianchinotti V. Molecular identification of three isolates of *Trichoderma harzianum* isolated from agricultural soils in Argentina, and their abilities to detoxify in vitro metsulfuron methyl. Botany. 2015;**93**(11):793-800. DOI: 10.1139/cjb-2015-0085

[87] Zafra G, Moreno-Montaño A, Absalón ÁE, Cortés-Espinosa DV. Degradation of polycyclic aromatic hydrocarbons in soil by a tolerant strain of *Trichoderma asperellum*. Environmental Science and Pollution Research. 2015;**22**:1034-1042. DOI: 10.1007/s11356-014-3357-y

[88] Sun J, Yuan X, Li Y, Wang X, Chen J. The pathway of 2, 2-dichlorovinyl dimethyl phosphate (DDVP) degradation by *Trichoderma atroviride* strain T23 and characterization of a paraoxonase-like enzyme. Applied Microbiology and Biotechnology. 2019;**103**:8947-8962. DOI: 10.1007/s00253-019-10136-2

[89] Sharma P, Sharma M, Raja M, Singh DV, Srivastava M. Use of *Trichoderma* spp. in biodegradation of carbendazim. Indian Journal of Agricultural Sciences. 2016;**86**(7):891-894

[90] Linhart C, Niedrist GH, Nagler M, Nagrani R, Temml V, Bardelli T, et al. Pesticide contamination and associated risk factors at public playgrounds near intensively managed apple and wine orchards. Environmental Sciences Europe. 2019;**31**:1-16. DOI: 10.1186/s12302-019-0206-0

[91] Abd-El-Kareem F, Elshahawy IE, Abd-Elgawad MM. Local *Trichoderma* strains as a control strategy of complex black root rot disease of strawberry in Egypt. Bulletin of the National Research Centre. 2019;**43**(1):1-7. DOI: 10.1186/s42269-019-0206-7

[92] Ahmed MFA, El-Fiki IAI. Effect of biological control of root rot diseases of strawberry using *Trichoderma* spp. Middle East Journal of Applied Sciences. 2017;**7**(3):482-492

[93] Morales-Mora LA, Andrade-Hoyos P, Valencia-delta M, Romero-Arenas O, Silva-Rojas HV, Contreras-Paredes CA. Characterization of strawberry associated fungi and in vitro antagonistic effect of *Trichoderma harzianum*. Revista Mexicana de Fitopatología. 2020;**38**(3):434-449. DOI: 10.18781/r.mex.fit.2005-7

[94] Mirmajlessi SM, Mänd M, Najdabbasi N, Larena Nistal MI, Loit E. Screening of native *Trichoderma harzianum* isolates for their ability to control Verticillium wilt of strawberry. Zemdirbyste-Agriculture. 2016;**103**(4):397-404

[95] Hannan A, Atta S, Bashir MA, Hyder S, Tauqir NA, Ali M, et al. Comparative efficacy of commercial fungicides and *Trichoderma harzianum* against Fusarium wilt in strawberry. In: Congress Book. Cyprus West University and Nigde Omer Halisdemir University. Turkey: Turkish Science and Technology Publishing; 25-29 Oct 2021. p. 265

[96] Lombardi N, Caira S, Troise AD, Scaloni A, Vitaglione P, Vinale F, et al. *Trichoderma* applications on strawberry plants modulate the physiological processes positively affecting fruit production and quality. Frontiers in Microbiology. 2020;**11**:1364. DOI: 10.3389/fmicb.2020.01364

[97] Sekmen Cetinel AH, Gokce A, Erdik E, Cetinel B, Cetinkaya N. The effect of *Trichoderma citrinoviride* treatment under salinity combined to *Rhizoctonia solani* infection in strawberry (*Fragaria* x *ananassa* duch.). Agronomy. 2021;**11**(8):1589. DOI: 10.3390/agronomy11081589

[98] Awad HM. Integrated management of Rhizoctonia Root rot disease infecting strawberry in Egypt. Egyptian Journal of Crop Protection. 2016;**11**(1):1-11

[99] Hu YJ, Yang HM, Jin J, Yan HH, Wang JP, Zhang RQ. Synergistic activity of antagonistic *Trichoderma* spp. and *Rhizoctonia solani* increases disease severity on strawberry petioles. European Journal of Plant Pathology. 2022;**164**(3):375-389. DOI: 10.1007/s10658-022-02568-w

[100] Abied MA, Khalil MI, Awad MF, Youssef SA. Alleviation of black root rot symptoms and alteration of strawberry growth via modulating physiological and biochemical mechanisms using *Trichoderma viride* and *Bacillus subtilis*. European Journal of Plant Pathology. 2023:1-16. DOI: 10.1007/s10658-023-02697-w

[101] Khalifa ESZ, Amer GAA, Bakr RA, Hamad AS. Bio-control efficacy of *Trichoderma* against strawberry charcoal rot disease. Egyptian Journal of Crop Protection. 2019;**14**(1):1-10

[102] Korkom Y, Yıldız A. Çilek üretim alanlarından izole edilen *Trichoderma* izolatlarının çilekte (cv. Rubygem) *Macrophomina phaseolina*'ya karşı etkinliğinin değerlendirilmesi. Adnan Menderes Üniversitesi Ziraat Fakültesi Dergisi. 2020;**17**(1):21-28. DOI: 10.25308/aduziraat.619308

[103] Korkom Y, Yildiz A. Evaluation of biocontrol potential of native *Trichoderma* isolates against charcoal rot of strawberry. Journal of Plant Pathology. 2022;**104**(2):671-682. DOI: 10.1007/s42161-022-01063-9

Section 3

Biochemical Contents of Berries

Chapter 6

Research of Antioxidant Activity of *Aronia melanocarpa* Fruits and *Viburnum opulus* Fruits

Valentina Golikova

Abstract

The general interest and the possibility of extensive use of polyphenol compounds determines the relevance of search of new low-cost and available sources of raw rnaQ-ial. In this aspect the *Aronia melanocarpa*. Viburnum op ulus are promising. I decided to determine the optimal concentration of ethyl alcohol, providing the highest antioxidant activity were prepared tinctures *A. melanocarpa* fruits, *Viburnum opulus* fruits and their combination. For this I took tinctures were made 1: 5 with percolation. Alcohol (40%, 70%, 95%) was used as an extracting agent; medicinal plant material - *V. opulus* fruits, *A. melanocarpa* fruits. Study of AOA was carried using test systems and method of EPR spectroscopy. Statistical analysis of the obtained data was done with Statgraphics (n = 6). Conclusion: The highest antioxidant activity was shown tincture *V. opulus* fruits 40%. In that way the optimal concentration of ethyl alcohol, providing the highest antioxidant activity is 40%.

Keywords: antioxidant activity, *Aronia melanocarpa* fruits, *Viburnum opulus* fruits, tincture, plant

1. Introduction

The development of modern medicine makes it possible to prevent and cure many diseases. At the same time, significantly increased the number of chronic diseases has greatly increased. In search of new effective and safe drugs, clinicians and pharmacologists turn to medicinal plants.

The organism has its own system of fighting excessive amounts of free radicals, but it is weakened by the adverse effects of the environment and needs constant support from outside. Many studies have proven that flavonoids are able to bind free radicals, therefore, exhibit antioxidant activity.

The "French paradox" that caused surprise made scientists pay attention to polyphenolic compounds of plants. Polyphenols, diverse in their structure, have a diverse pharmacological effect on the human body.

The general interest and the possibility of extensive use of polyphenol compounds determines the relevance of search of new low-cost and available sources of raw material. In this aspect the *A. melanocarpa*. *V. opulus* are promising [1].

1.1 *A. melanocarpa*

Aronia prune - *A. melanocarpa* (Michx.) Elliott., family of *Rosaceae - Rosaceae.* A few decades ago, few people knew about Aronia prune. Since 1843, aronia has been grown only in botanical gardens. However, local residents of Altai began to taste the fruits of an unknown plant, as a result of which many began to feel much better: headaches stopped, gastrointestinal disorders passed, sleep improved. These observations formed the basis for a detailed study of the chemical composition of berries and the determination of their therapeutic effect in the clinic, which began in 1953.

A. melanocarpa is a small shrub up to 2.5 m high. Shoots are numerous with simple whole leaves obovate and saw-edged; leaves are dark green, turning red in autumn. The flowers are regular, five-membered, white or slightly pink, collected in corymbose inflorescences **Figure 1**.

1.2 The main groups of biologically active substances of *A. melanocarpa*

Aronia fruits contain a P-vitamin complex consisting of flavonoids: rutin, catechins, quercetin, hesperidin, anthocyanins, as well as a significant amount of ascorbic acid (up to 110 mg%).

Other groups of BAS:

1. tannins (upto 0.6%),

2. organic acids,

3. sugars (fructose, glucose, etc. (up to 10%)),

4. glucoside-amygdalin (5–30 mg),

5. pectin substances (up to 0.75%),

6. vitamins of group B, A, C (up to 100 mg%).

Figure 1.
Appearance of Aronia melanocarpa.

Of inorganic substances, fruits contain:

Aronia is distinguished by a large set of trace elements — it contains boron, fluorine, iodide compounds (6–10 micrograms per 100 g of fresh fruit), iron, copper, manganese, molybdenum. Se accumulates.

From the above, it can be concluded that the chemical composition of various parts of the plant is quite diverse, but the main class of biologically active substances contained are flavonoids [2].

1.3 Pharmacological properties

Aronia lowers cholesterol in the blood of patients with atherosclerosis. Fresh, dry, pureed fruits with sugar and juice of aronia are used for therapeutic and prophylactic purposes in various conditions accompanied by increased permeability of blood vessels — glomerulonephritis, radiation damage, rheumatism, allergies, measles, typhus, scarlet fever.

The fruits of aronia contain sorbitol, which is important for the nutrition of patients with diabetes mellitus, with diseases of the liver and biliary tract. Fruits and juice of aronia, which preserve vitamins well, increase the body's defenses, increase appetite, increase the digestibility and acidity of gastric juice.

Fresh fruits are used as a vitamin remedy for hypo- and vitamin deficiency P and hypertension of stages 1 and 11.

The juice helps strengthen the walls of blood vessels. The leaves contain substances that improve liver function, the formation and outflow of bile [3].

Recently, substances have been identified in aronia that increase blood pressure to normal, if it was reduced. P-active compounds of aronia fruits have the ability to bind and remove radioactive strontium from the body. Preparations from aronia have vasodilating properties.

Among many doctors, there was a fear that the use of aronia would cause increased thrombosis in the human body. This fear was based on the fact that aronia contains a large number of substances with P-vitamin properties. However, experimental and clinical studies have shown that aronia affects blood clotting only in cases when it is lowered, that is, with bleeding.

1.4 *Viburnum opulus*

V. opulus is a small branching shrub or a small tree 1.5–4 m high with graybrown bark. The leaves are petiolate, opposite, broadly ovate or rounded, 5—8 cm long and 5–8 cm wide, three-five-lobed **Figure 2** [1].

The main groups of biologically active substances of viburnum vulgaris: the fruits contain carbohydrates: fructose, glucose, sucrose, mannose, galactose, xylose, arabinose, polysaccharides; pectin substances, organic acids: aceticand isovaleric (up to 3%); triterpenoids: oleonolic and hederagenic acids and their acetyl derivatives, ursolic acid; steroids, vitamin C (up to 0.09%) and carotene; phenol-carboxylic acids and their derivatives. Also flavonoids, tannins, catechins.

1.5 Pharmacological properties of *Viburnum opulus*

The infusion is used as a vitamin, diaphoretic, diuretic and disinfectant. Liquid extract and decoction, used as a hemostatic agent, mainly for uterine bleeding.

Figure 2.
Appearance of Viburnum opulus.

Flavonoids are a numerous group of phenolic compounds, the structure of which is based on a skeleton consisting of 2 benzene rings connected by a three-carbon chain [4].

1.6 Pharmacological action of flavonoids

Most flavonoids have high P-vitamin activity, that is, they are able to reduce the fragility and permeability of capillary walls. Currently, preparations with antiinflammatory and anti-ulcer effects, as well as choleretic and hepatoprotective have been obtained on the basis of flavonoids. As a result of recent studies, hypoglycemic and antiviral drugs have been obtained.

Flavonoids have an antispasmodic effect. Flavonols mainly affect the neutralizing function of the liver, the mechanism of action is associated with changes in redox processes in the mitochondria of liver cells.

Isoflavonoids have a moderate estrogen-like effect, they are sometimes referred to as phytoestrogens. Raw materials containing these compounds (species of the genus clover) are used in the production of biologically active additives, which are used in gynecology for various hormonal disorders.

Thus, it can be concluded that flavonoids have versatile pharmacological activity, which, with low toxicity, makes them a very valuable and promising group of biologically active substances [1, 2].

1.7 Antioxidant activity of plant material

Antioxidants are substances that slow down or prevent the oxidation of organic compounds. They protect our body from the bad effects of free radicals. The antioxidant creates a barrier to the destructive action of an additional electron when it combines with a free radical. The body converts the cellular oxidant into water and oxygen with the help of an enzymatic defense system.

Oxygen is a powerful oxidizer, oxidation reactions with its participation are a source of energy for many living organisms. On the other hand, oxygen compounds are formed during metabolism, which destroy the cell structure. As a result, metabolism is disrupted not only in the cell, but throughout the body. The role of antioxidants is to bind and remove free radicals from the body.

The body has its own system for combating excessive amounts of free radicals, but it is weakened due to exposure to polluted environments, smoking, direct sunlight and needs support. It was found that many plants contain substances flavonoids - a large group of compounds with a polyphenolic structure that bind free radicals, that is, they are antioxidants [5].

1.8 The role of free radicals in a living organism

A free radical is an atom or group of atoms having an unpaired electron at the last electronic level, which makes them extremely unstable. In this state, free radicals trap vulnerable proteins, enzymes, lipids, and even whole cells. By taking an electron from a molecule, they inactivate cells, thereby disrupting the fragile chemical balance of the body. When the process happens again and again, a chain reaction of free radicals begins, while cell membranes are destroyed, important biological processes are undermined, mutant cells are created. Free radicals can reversibly or irreversibly destroy substances of all biochemical classes, including free amino acids, lipids, carbohydrates and connective tissue molecules [6].

Over the past few years, it has been proven that antioxidants are extremely useful for the body - they prevent the development of cardiovascular diseases, protect against cancer and premature aging, also increase immunity and much more. The last decade has provided a lot of evidence proving that free radicals play a role in the development of many diseases. If free radicals oxidize lipids, a dangerous form of lipid peroxide is formed. Many scientists associate the formation of lipid peroxides with cancer, heart disease, accelerated aging and immune deficiency.

Free radicals are constantly formed in the body, so there must be antioxidant protection from them, which is one of the most important components of immunity in general. It is important to supplement your diet with natural substances - antioxidants, which enhance protection against free radicals, thereby increase immunity, the body's resistance to adverse external factors, slow down the aging process [5].

The most important antioxidants are: vitamins C, E, beta-carotene, selenium, bioflavonoids (vitamin-like substances contained in the skin of plants - oranges, lemons, tomatoes, etc.). Many plant extracts, vitamins, amino acids, minerals, trace elements have antioxidant properties either directly or indirectly, as they are part of antioxidant enzymes.

About 5000 flavonoids - antioxidants with a wide range of healing effects were found in plants. They have vasodilating, antitumor, anti-inflammatory, bactericidal, immunostimulating and antiallergic properties.

Any organism can be considered as an example of a balanced and well-functioning antioxidant system consisting of many components - vitamins (vitamins C, E, P), enzymes and trace elements (selenium, zinc), polyphenolic compounds (flavonoids), and sulfurcontaining amino acids (cysteine, methionine), as well as the tripeptide glutathione. These are only some compounds that have an antioxidant effect. The chemical nature of these compounds is diverse, among them there are both water- and fat-soluble components. The main principle on which the action of the antioxidant system of a living organism is based is synergy. It consists in the fact that the components of the system work together, restoring each other and enhancing the effectiveness of the action [6].

In the course of the work, research objects and auxiliary substances were used, which corresponded in terms of qualitative indicators and quantitative content to the requirements of regulatory documentation.

2. Material and methods

Tinctures were made 1:5 with percolation. Alcohol (40%, 70%, 95%) was used as an extracting agent; medicinal plant material - *V. opulus* fruits, *Aronia melanocarpa* fruits. Study of AOA was carried using test systems and method of EPR spectroscopy. Statistical analysis of the obtained data was done with Statgraphics (n = 6).

Tinctures were obtained in a ratio of 1:5 by percolation, extractant - 40%, 70%, 95% ethyl alcohol. As medicinal plant raw materials used: *V. opulus* fructus, *A. melanocarpa* fructus, and their combination (0.5,0.5). The total content of flavonoids in terms of catechin was carried out as follows: for the preparation of the standard, 20 mg of catechin was taken and placed in a 50 ml flask, brought to the mark with purified water. 1.4 ml of purified water and 100 ml of the test sample, 60 ml of 5% sodium nitrite solution and 60 ml of 10% aluminum chloride solution were placed in the cuvette and mixed well. To prepare a reference sample, 1.4 ml of water, 100 ml of catechin solution, 60 ml of 5% sodium nitrite solution and 60 ml of 10% aluminum chloride solution are placed in a cuvette. They were kept for 5 minutes in a thermostat at 25°C, then 0.4 ml of IM NaOH was added, mixed well and the optical density was measured at 510 nm. Further calculations were carried out according to the calibration schedule [7].

The antioxidant activity was determined using the free radical 1,1-diphenyl-l-picrylhydrazyl (DPPG), using electron paramagnetic resonance (EPR). This method is based on the determination of changes in the concentration of DPH as a result of a reaction with an antioxidant. EPR operates under the influence of a magnetic field with a constant frequency (microwave). The EPR spectrometer at the Faculty of Pharmacy of the Warsaw Medical University operates in the X band (microwave 9.4 GHz) [8].

Dilute 50 ml of each tincture with water. Add a DFPG of about 1.3 mm. Stir on a Vortex machine for 30 seconds. The solution is kept for 20 minutes in a dark place and re-mixed. The capillary is filled with the resulting solution and EPR spectra are recorded [9].

3. Research results and their discussion

The content of flavonoids (in terms of catechin) in tinctures (mg KA/ml): Aroniae melanocarpae + Viburni opuli 40% - 5.9766 ± 0.455576; Aroniae melanocarpae 70% - 1.00667 ± 0.253739; Aroniae melanocarpae + Viburni opuli 95% - 2.07 + 0.0496828; Aroniae melanocarpae 70% - 548,667 ± 0,339,095; Aroniae melanocarpae 95% - 0.713333 ± 0.0143422; Aroniae melanocarpae 40% - 3.37 ± 0.430265; Viburni opuli 70% - 5.17333 + 0.539883; Viburni opuli 40% - 6.00333 + 0.6212; Viburni opuli 95% - 2.26333 ± 0.248827 **Figure 3**. Statistical processing in appendix 1 **Tables 1–3**.

Thus, the following conclusions can be drawn:

1. The quantitative content of flavonoids in tinctures based on *Viburnum opulus* fructus, Aroniae melanocarpae and their combinations were determined;

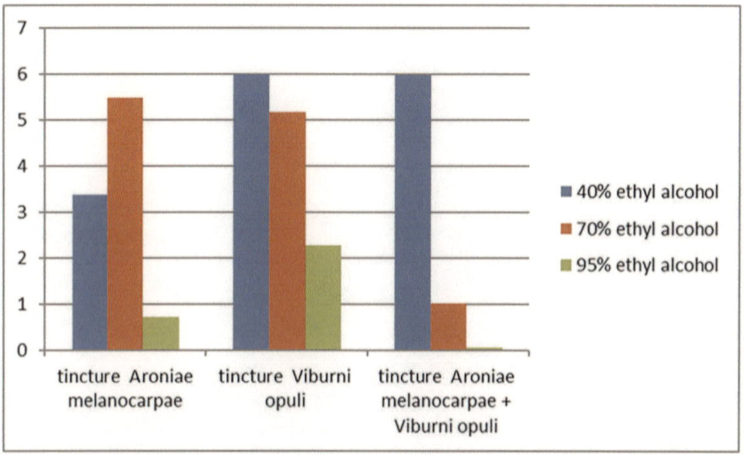

Figure 3.
The content of flavonoids in the studied tinctures.

2. The optimal alcohol concentration was determined, which provides the highest yield of flavonoids. Such concentrations are 40% and 70%:

   - In tinctures based on *Aronia melanocarpa* fructus, the largest amount of flavonoids is contained in 70%concentration and is 5.48667 ± 0.339095;

   - In a tincture based on *V. opulus* fructus, the highest content of flavonoids is manifested in 40% concentration and is 6.00333 ± 0.6212;

   - In the combined tincture, the highest content of flavonoids is observed in 40% concentration and is 5.9766 ± 0.455576.

3. It was found that the largest amount of flavonoids contains 40% tincture of 1 krnoaddU-JClJS (16.00333 ± 0.6212).

   Antioxidant activity (mg/ml): Aroniae melanocarpae 40% - 44.63 ± 0.08618, Aroniae melanocarpae 70% - 13.86 ± 0.06992, Aroniae melanocarpae 95% - 18.09 ± 0.02548, Aroniae melanocarpae + Viburni opuli 40% - 57.76 ± 0.08455; Aroniae melanocarpae + Viburni opuli 70% - 71.39 ± 0.3164; Aroniae melanocarpae + Viburni opuli 95%- 24.55 ± 0.07154, Viburni opuli 40% - 162.97 ± 0.0322; Viburni opuli 70% - 158.06 ± 0.0186; Viburni opuli 95%-37.06 ± 0.08178 **Figure 4**. Statistical processing in appendix 2 **Tables 4–6**.

   The results of the study were processed statistically using STATGRAPHICS Plus for Windows (n = 6).

Thus, the following conclusions can be drawn:

1. Antioxidant activity was determined in tinctures based on viburnum vulgaris, aronia melanocarpus and their combinations;

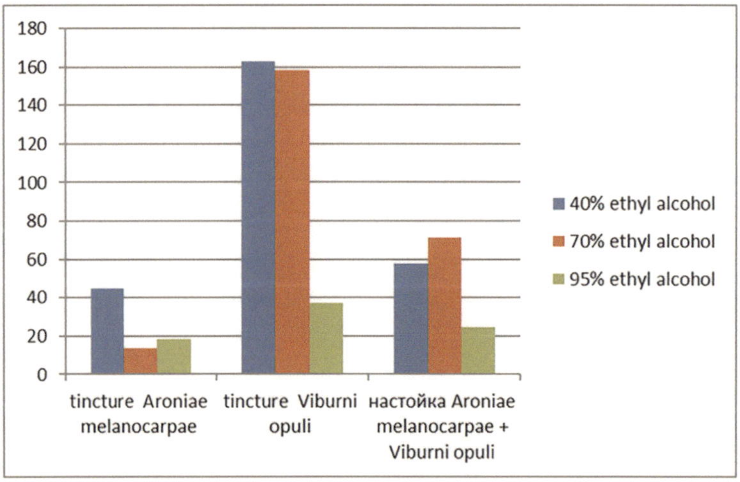

Figure 4.
Results of antioxidant activity in the studied tinctures.

2. Having determined at what concentration of alcohol the greatest antioxidant activity is manifested, we have obtained that:

   - In tinctures based on *Aronia melanocarpa* fructus, the greatest antioxidant activity is manifested in 40% concentration and amounts to 44.63 ± 0.08618;

   - In the tincture based on viburnum ordinary fruit, the greatest antioxidant activity is manifested in 40% concentration and is 162.97 ± 0.0322;

   - In the combined tincture, the greatest antioxidant activity is manifested in 70% concentration and is 71.39 ± 0.3164.

3. It was found that the greatest antioxidant activity is shown by 40% tincture of viburnum vulgaris (162.97 ± 0.0322).

4. Conclusion

The highest antioxidant activity was shown tincture *Viburnum opulus* fruits 40%. In that way the optimal concentration of ethyl alcohol, providing the highest antioxidant activity is 40%.

A. Appendices 1

See **Tables 1–3**.

Raw material	Alcohol concentration (%)	Optical density	Content of flavonoids (divided), mg KA/ml	Content of flavonoids (concentrated), mg KA/ml	Average
A. melanocarpa fructus	40	0.220	0.16	3.47	3.37 ± 0.430265
		0.241	0.17	2.47	
		0.241	0.17	3.47	
	70	0.073	0.005	0.88	5.48667 ± 0.339095
		0.060	0.04	0.89	

Table 1.
Results of determining the quantitative content of flavonoids in the tincture of Aronia melanocarpa *fructus.*

Raw material	Alcohol concentration (%)	Optical density	Content of flavonoids (divided), mg KA/ml	Content of flavonoids (concentrated), mg KA/ml	Average
V. opulus fructus	40	0.220	0.16	3.17	6.00333 ± 0.6212
		0.241	0.17	3.47	
		0.241	0.17	3.47	
	70	0.364	0.26	5.37	5.17333 ± 0.539883
		0.375	0.27	5.37	
		0.345	0.25	4.94	
	95	0.149	0.11	2.16	2.26333 ± 0.248827
		0.157	0.11	2.27	
		0.163	0.12	2.36	

Table 2.
Results of determining the quantitative content of flavonoids in the tincture of Viburnum opulus *fructus.*

Raw material	Alcohol concentration (%)	Optical density	Content of flavonoids (divided), mg KA/ml KA/мл	Content of flavonoids (concentrated), mg KA/ml	Average
Aroniae melanocarpae + Viburni opuli fructus	40	0.428	0.31	6.12	5.9766 ± 0.455576
		0.422	0.30	6.04	
		0.403	0.29	5.77	
	70	0.388	0.28	5.55	5.48667 ± 0.339095
		0.372	0.27	5.33	
		0.390	0.28	5.58	
	95	0.144	0.10	2.09	2.07 ± 0.0496828
		0.141	0.10	2.05	
		0.143	0.10	2.07	

Table 3.
Results of determining the quantitative content of flavonoids in the tincture of Aroniae melanocarpae + Viburni opuli *fructus.*

B. Appendices 2

See **Tables 4–6**.

Raw material	Alcohol concentration (%)	DDPH (mg/ml), divided	DDPH (mg/ml), concentrated	Average
A. melanocarpa fructus	40	2.07	41.47	44.63 ± 0.08618
		2.11	42.29	
		2.51	50.12	
	70	0.28	5.65	13.86 ± 0.06992
		1.02	20.44	
		0.77	15.48	
	95	0.09	1.85	8.09 ± 0.02548
		2.09	41.85	
		0.53	10.56	

Table 4.
Results of determination of antioxidant activity in tincture of Aronia melanocarpa *fructus.*

Raw material	Alcohol concentration (%)	DDPH (мг/мл), divided	DDPH (мг/мл), concentrated	Average
V. opulus fructus	40	4.14	206.89	162.97 ± 0.0322
		3.09	154.57	
		2.55	127.44	
	70	0.98	68.71	158.06 ± 0.0186
		4.21	294.82	
		1.58	110.66	
	95	0.30	14.88	37.06 ± 0.08178
		0.96	47.96	
		0.97	48.34	

Table 5.
Results of determination of antioxidant activity in tincture of Viburnum opulus *fructus.*

Raw material	Alcohol concentration (%)	DDPH (мг/мл), divided	DDPH (мг/мл), concentrated	Average
Aroniae melanocarpae + *Viburni opuli* fructus	40	2.96	59.18	57.76 ± 0.08455
		2.38	47.53	
		3.33	66.57	
	70	3.10	62.08	71.39 ± 0.3164
		3.88	77.64	
		3.72	74.45	
	95	1.11	22.19	24.55 ± 0.07154

Table 6.
Results of determination of antioxidant activity in tincture of Aroniae melanocarpae + Viburni opuli *fructus.*

Author details

Valentina Golikova
Smolensk State Medical University, Smolensk, Russia

*Address all correspondence to: zeleneeva.valentina@yandex.ru

References

[1] Kukes VG, Grachev SV, Sychev DA, Ramenskaya GV. Drug metabolism. In: Scientific Foundations of Personalized Medicine: A Guide for Doctors. Moscow: GEOTAR-Media; 2008. p. 304

[2] Kurkin VA. Pharmacognosy. In: Textbook for Pharmaceutical Universities and Faculties. Samara: SamSMU; 2004. p. 1239

[3] Ryazanova TK. Pharmacognostic study of fruits and shoots of blueberries. Fundamental Research. 2013;**8–5**: 1136-1140

[4] Lotova LI. Morphology and Anatomy of Higher Plants. Moscow: Editorial URSS; 2000. p. 528

[5] Abdulin IF, Turova EN, Budnikov GK. Organic antioxidants as objects of analysis. Factory laboratory. Diagnostics of materials. 2001;**167**(6): 3-13

[6] Ermakova VA, Sorokina AA, Nesterova OV. Medicinal Plants and Medicinal Plant Raw Materials Containingtannins: Studies, Methodical, Manual. Moscow: Russian doctor; 2005. p. 155

[7] Astakhova AV, Lepakhin VK, Medicines. Adverse Adverse Reactions and Safety Control. 2nd ed, ispr. and add. Moscow: Eksmo; 2008. p. 256

[8] Bondet V, Brand-Williams W, Berset C. Kinetics and mechanisms of antioxidant activity using the DPPH* free radical method. Lebensmittel-Wissenschaft und - Technologie. 1997; **30**:609-615

[9] Deineka LA, Blinova LP, Chulkov AN, et al. Method of extraction and purification of anthocyanins from fruits of aronia chernoplodnaya. Scientific Vedomosti BelSU. 2012;**10** (129) issue 18/2:60-65

Chapter 7

Mulberry

Zoha Sohail

Abstract

Mulberry plant belongs to the family Moracea from the genus *Morus* and can be cultivated in mountains, plains, and valleys, even in humid and rain-fed areas. Many various varieties of this plant are present around the world, but three prime varieties include, that is, native red mulberry (*Morus rubra*), East Asian white mulberry (*Morus alba*), and southwestern Asian black mulberry (*Morus nigra*). The presence of certain biologically active compounds in its fruit provides several pharmacological benefits to health. Polyphenols, flavonols, anthocyanin, flavonoids, anthocyanins, benzoic acid, and hydroxycinnamic acid are the major bioactive active compounds responsible for pharmacological benefits, that is, anticholesterol, antidiabetic, antioxidative, and antiobesity effects fruit of this plant can be consumed as whole fruit, as juice, and as spread but its preservation is difficult due to its high water content. Therefore, this chapter will aim to focus on different types of mulberry plants and the pharmacologic benefits of different mulberry fruits.

Keywords: mulberry, *Morus alba*, *Morus rubra*, *Morus nigra*, biological active compounds, pharmacologic benefits

1. Introduction

The family Moraceae shelters mulberry (*Morus*), an East Asian plant; throughout the world, this plant is widely spread in various biological and geographic zones by farming in tropical, subtropical, and temperate climates and by natural occurrence in forestry [1]. The word "mulberry" has a stimulating etymological progression; the German word for berry, beri, was combined with the Latin word *Morus* to create the word múlbere, múlbere was then transformed into the German name Maulbeer, afterward which was then morphed into the Dutch word moerbezie, and ultimately was transformed to mulberry an English word [2]. Its origin is in Asia, grows in length to about 16 m, an ecologically important plant as its leaves are used as feed for silkworms and this plant is wind cross-fertilized and also set fruit without pollination [3]. Depending on the plant's age and cultivars, mulberry fruit contains a wide diversity of nutritive and bioactive compounds, that is, amino acids, fatty acids, vitamins, minerals, rutin, quercetin, anthocyanins, chlorogenic acid, and polysaccharides. Due to the presence of these compounds in its fruit, it possesses antioxidative, neuroprotective, antiatherosclerotic, immunomodulative, antitumor, antihyperglycemic, and hypolipidemic properties [4] among berries; mulberries are richest in phenolics and anthocyanins [5]. Mulberry fruits are categorized as functional foods owing to the occurrence of these bioactive chemicals and the provision of medicinal properties important for human health [6]. Although this fruit is native to Asia but has been broadly spread to Europe, Asia, North America, Africa, Latin America, and South America [7]. Three main types of this plant include white mulberry (*M. alba*) originated in Western Asia, red mulberry (*Morus rubra*) originated in North and South America and black mulberry (*Morus nigra*) originated from Southern Russia [8]. The Mulberry plant offers multibenefits, that is, food, fodder, and fuel; its fruit is eaten as a raw fruit as fruit juice as fruit puree, and also used in sericulture [9] used as molasses, vinegar, and jam [10]. Communities in southeast region consume dried mulberries as an appetizer. They process mulberry in its juice and then combine the juice with starch afterward, boiling and drying under the sun, forming pestil; they also mix mulberry fruit juice with starch and walnut kernel, forming a solid sausage shape and lastly, drying under the sun forming kome [11] thus providing substantial marketing value due to the nutritious and distinct characteristics [10]. It has been affirmed that mulberry fruits and their products, such as molasses, juices, compotes, and jams, could aid in the management of some coronary, gastric, and intestinal problems [12]. Agrarians in China and India cultivate mulberry plants for silkworms, and European cultivate mulberry plants for their fruit [13]. With respect to fruit production, there are three prime varieties of mulberries, that is, *M. alba*, *Morus rubra*, and *Morus nigra*.

2. Major types of mulberry fruit

2.1 White mulberry (*M. alba*)

M. alba plant is also known as Tutam in Sanskrit, Tut in Hindi, Mulberry in English, Malbari in Malayalam, and Musukette in Tamil. This plant is moderately sized, three to six meters tall, and categorized as a shrub or tree. In Chinese medicine, *M. alba* plant provides a long list of medicinal benefits and traditionally its fruit is traditionally used to nourish blood, treat anemia, fatigue, and prevent premature graying of hair. Besides its fruit, all its parts offer medicinal benefits in treating asthma, edema, cough, bronchitis, insomnia, influenza, diabetes, eye infections, and nose bleeding. The indigeneous system has providediuretic, laxative, antibacterial, and anthelmintic properties [14].

2.2 Red mulberry (*M. rubra*)

Morus is a Latin word for mulberry, and rubra is a Latin word for red. Trees of this plant are found in damp soils, floodplains, and moist sites in South Florida, West Texas, Minnesota, Ontario, and mid-Atlantic regions. In different areas, different parts of this plant are used to treat diseases, that is, Albama and Indians are using its root to treat urinary tract issues, its juice is being used by Rappahannock community to treat ringworm by rubbing over the skin, Cherokee uses its bark to eliminate intestinal worms by steeping its bark and ingesting the liquid lastly, its root bark is used in Meskwaki to treat many ailments [2].

2.3 Black mulberry (*M. nigra*)

M. nigra is a species of plant in the family Moraceae, innate in humid, subtropical, and pleasant zones in Asia, Europe, North America, South America, and Africa [13]. This is one of the very important specie of *Morus* genus, its sensory evaluation indicates that it has sweet, sour, musky and woody flavors and has woody, fresh aroma physiognomies [15], dark black color of the fruit due to the presence of anthocyanins [16]. This plant is grown in many countries without any particular protection, and fortunately, its barks, leaves, and berries all provide medicinal properties, that is, berries for treating inflammation and blood bleeding, bark for treating toothache, and leaves are used as an antidote. Fortunately, its trees are super resistant to pests and other diseases, due to which its fruit can make products with good nutritional value leading to healthy, organic, and ecological lifestyle [17]. In Europe, leaves of *M. nigra* have been used to stimulate the production of insulin production in people with diabetes mellitus [18]. In Turkey, *M. nigra* has been used in folk medicine to treat fever, to buffer the liver, to support joints, to manage blood pressure, and to assist the passage of urine [19].

3. Nutritional composition of mulberry fruit

Mulberry fruit, when ripened, offers great taste, aroma, and flavor. It is valued to consume it as raw fruit rather than consuming it in its processed form [20]. The nutritional composition of fruits is dependent on the ripeness, type of soil, season, variants in genotypes, and postharvest observations. Taste, color, and browning of the fruit are dependent on the concentration of phenolics compounds [21]. Mulberry fruits contain good water content, that is, >70%, are low in calories, and are rich in nutrients and antioxidants; they have great composition to maintain a healthy body status [22]. Mulberries are an excellent combination of polyphenols, that is, flavones, isoflavanols, lignans, tannins, anthocyanins, catechins, and flavonols. The popularity of this fruit in the human fruit is basically due to the presence of these phytochemicals [23].

Mulberries grown in the Xinjiang region of China have high potassium content, followed by calcium and magnesium, the high iron content of black mulberry from China has made them a valuable source for overcoming iron deficiencies. A high percentage of linoleic acid, that is, 52.3% present in white mulberry obtained from Xinjiang, China [24]. 4.2 mg/100 g iron is present in *M. alba* and nigra, Russian mulberry and black mulberry obtained from China have high iron content ranging from 11.4 to 11.9 mg/100 g FW, and the bioavailability of iron can be enhanced by vitamin C's chelating properties; the highest value of vitamin C is present in *M. alba* (22.4 mg/100 ml) followed by *M. nigra* (21.8 mg/100 ml) and then *M. rubra* (19.4 mg/100 ml) [22]. In context to the fatty acid composition of mulberry fruits, linoleic acid is the most dominant one, followed by palmitic and oleic acid [22]. Mulberry fruits obtained from southeastern Spain contain a good percentage of protein and are considered as a high protein source; *M. alba* has high protein content than *M. nigra* [25].

3.1 Major phytochemicals

Anthocyanins: A flavonoid and a major contributor to the color of flower and fruit ranging from red to blue and purple, gained great importance due to its biological benefits, that is, antitumor, antidiabetic, and antioxidant properties. Mulberry fruit contains many various types of anthocyanins, that is, cyanidin-3-glucoside (C3G), cyanidin-3-rutinoside (C3R), pelargonidin-3-glucoside, pelargonidin-3-rutinoside, cyanidin 3-O-(6"-O-α-rhamnopyranosyl-β-D-glucopyranoside), cyanidin 3-O-(6"-O-α-rhamnopyranosyl-β-D-galactopyranoside), cyanidin 3-O-β-D-galactopyranoside, cyanidin 7-O-β-D-glucopyranoside, and petunidin 3-O-β-glucopyranoside. Among these all, the most principle one in mulberry fruit is C3G and then C3R, also contains minimal amounts of pelargonidin-3-glucoside and pelargonidin-3-rutinoside, these two compounds, that is, cyanidin 3-O-(6'-O-α-rhamnopyranosyl-β-D-glucopyranoside) known as keracyanin and cyanidin 3-O-(6'-O-α-rhamnopyranosyl-β-D-galactopyranoside) have been identified by high-speed counter current chromatography [4].

Flavonols and flavanols: flavonols present in mulberry fruit include rutin, quercetin, quercetin 3-O-rutinoside, quercetin 3-oglucoside, quercetin 3-O-galactoside, myricetin, kaempferol, kaempferol 3-O-glucoside, and kaempferol3-O-rutinoside and flavanols present in it include catechin, epigallocatechin gallate, epicatechin, procyanidin B1, and procyanidin B2. Among all these, rutin is the most abundant one flavonols, quercetin, and kaempferol's derivatives are its major components and their glycosylated forms, that is, quercetin 3-O-glucoside, quercetin 3-O-galactoside, and kaempferol3-O-rutinoside are found in some mulberry fruits [4].

Phenolic compounds and alkaloids: hydroxycinnamic acid and benzoic acid are two basic phenolic compounds in mulberry fruit. Compounds in hydroxycinnamic acid include chlorogenic acid, ferulic acid, *p*-coumaric acid, *o*-coumaric acid, and caffeic acid compounds in benzoic acid include gallic acid, *p*-hydroxybenzoic acid, syringic acid, protocatechuic acid, and vanillic acid. Among all these, chlorogenic acid occurs in the highest amount in mulberry, that is, 5.3–17.3 mg/100 g DW. Other major phenolic compounds include gallic acid, that is, 7.33–23.34 mg/100 g DW and cinnamic acid, that is, 11.64–15.05 mg/100 g DW. The concentration of phenolic compounds in different mulberries varies as geographical area and environmental factors, that is, temperature, light, humidity, and degree of maturity [4]. Alkaloids are pharmacologically active, therapeutic, and nitrogen-containing compounds [26]. In white mulberry, through NMR Spectroscopy, five alkaloids have been found, that is (morrole B, C, D, E, and F), and alkaloids present in other mulberry fruits include 2-formyl-5-(hydroxymethyl)-1H-pyrrole-1-butanoic acid, 5-(hydroxymethyl)-1H-pyrrole-2 carboxaldehyde, 2-formyl-1H-pyrrole-1-butanoic acid, and 2-formyl-5-(methoxymethyl)-1H-pyrrole-1-butanoic acid [27].

Aromatic and volatile compounds: Aromatic compounds, that is, benzaldehyde, ethyl butanoate, (E)-2-nonenal, 3-mercaptoethanols, and 1 hexanol are present in mulberry fruits [28] and volatile compounds present in white and black mulberry fruits include 2,4-nonadienal, methyl hexanate, limonene, octanol, and ethyl hexanoate [29].

Melatonin: A neurohormone also known as N-acetyl-5-methoxytryptamine, produced by the pineal gland, has a role in regulating circadian rhythms and sleep ailments and also possesses antioxidative and anti-inflammatory properties. Many fruits including mulberry have a good percentage, but the percentage differs as per fruit development, that is, high levels (5.76 ng/g FW) present in stage 1 and then decreased in stage 2 and 3 [4].

3.2 Nutrient composition of three major types

Mulberries are a great combination of bioactive compounds, that is, sugars (glucose and fructose), vitamins, minerals, fats (linoleic, palmitic, and oleic acid), and phenolic compounds, that is, alkaloids, flavonoids, tannins, and carotenoids [30].

M. alba: *M. alba* is a great composite of carbohydrates, lipids, proteins, vitamins, minerals, and fiber. The percentage of protein in *M. alba* is higher as compared to raspberries, strawberries, and blackberries [31]; it contains all essential amino acids, particularly branched-chain amino acids [21]; other than essential amino acids, *M. alba* also contains nonessential amino acids also its amino acid/total amino acid content is 42% that is equivalent to a good protein source, that is, fish and milk [32]. Macro and micronutrient composition of *M. alba* show that it contains 1.55 g/100 g dw protein, 0.48 g/100 g dw lipids, 1.47 g/100 g dw crude fiber, 0.57 g/100 mg dw ash content, 14.21 g/100 g dw total carbohydrates, moisture of fresh weight (fw) is 81.72 g/100 g dw, 7.55 g/100 g fw total sugar content, 0.088 mg/100 g fw riboflavin, 15.2 mg/100 g fw ascorbic acid, 0.19–0.37 g/100 g dw calcium, 0.24–0.31 g/100 g dw phosphorus, 1.62–2.13 g/100 g dw potassium, 0.12–0.19 g/100 g dw magnesium, 28.2–46.74 mg/kg dw iron, 4.22–6.38 mg/kg dw copper, 12.33–19.38 mg/kg dw manganese, 14.89–19.58 mg/kg dw zinc, 1.40–2,62 mg/kg dw nickel [33]. *M. alba* seeds are enriched with omega 3 fatty acids [34]. Upon analysis, its quoted that it contains the highest percentage of total fats, that is, 1.10%, followed by 0.95% in *Morus nigra* and then 0.85% in *M. rubra* [35] fatty acids composition shows that it contains linoleic acid and oleic acids in good amount along with this also contains trace quantities of eicosadienoic, linolenic, stearidonic, eicosanoic, and erucic acid [33]. The concentration of polyunsaturated fatty acids is highest in its fruit, that is, 76.68%, then monounsaturated fatty acids and lastly, saturated fatty acids [36]. Amount of total phenols is 0.13–0.40 mg GAE/g (fw), total flavonoids 29 mg/QE/100 g (fw), anthocyanins 0.36 mg/100 g (dw), carotenoids 13.16 mg/100 g, total sugars 7.55 g/100 g (fw), fructose 3.0–6.27 g/100 g (fw) and glucose 3.1–6.68 g/100 g (fw) [37]. Phytochemical investigation shows that it contains 5 anthocyanins (phenolic compounds serving as cancer-protective, cardioprotective, and inflammation protective agents) and 25 phenolic compounds [38].

M. nigra: It has been used for medicinal purposes since ancient times because of its good compositional status comprising both nutritional and biologically active nutrients, that is, carbohydrates (polysaccharides), proteins, vitamins, minerals, and polyphenols, that is, anthocyanin and quercetin. Furthermore, extracts of this fruit also attain anti-Alzheimer, antimicrobial, antitumor, and anticancerous qualities [39]. Amount of moisture 82.40% (fw), ash 0.50% (dw), carotenoids 14.0 mg/100 g (dw), protein 0.96 g/100 g (dw), fat 0.95 g/100 g (fw), fiber 11.75 g/100 g (dw), total phenols 0.42–0.80 mg GAE/g (fw), total flavonoids 276 mg QE/100 g (fw), total sugars 6.64 g/100 g (fw), fructose 1.5–5.63 g/100 g (fw), glucose 1.4–7.75 g/100 g (fw), anthocyanins 0.57 mg/100 g (fw), riboflavin 0.04 mg/100 g (fw), niacin 1.6 mg/100 g (fw), ascorbic acid 11.30–15.37 mg/100 g (fw), iron 4.2–42.13 mg/100 g (fw), calcium 132 mg/100 g (fw), phosphorous 232 mg/100 g (fw), sodium 59 mg/100 g (fw), potassium 922 mg/100 g (fw), magnesium 106 mg/100 g (fw), and nickel 0.27 mg/100 g (fw) [37].

M. rubra: Its nutritional composition analysis shows that it contains 1.26 g/100 g (fw) protein, 0.85 g/100 g(fw) fat, 3.2–5.41 g/100 g (fw) fructose, 3.3–6.07 g/100 g (fw) glucose, 19.38 mg/100 g (fw) anthocyanins, 16.17 mg/100 g (fw) ascorbic acid, 4.5–57.38, 132 mg/100 g (fw) calcium, 226 mg/100 g (fw) phosphorous, 61 mg/100 g (fw) sodium, 834 mg/100 g potassium, 115 mg/100 g magnesium, 3.2–5.04 mg/100 g (fw) zinc, 0.37 mg/100 g (fw) nickel, 0.20–0.90 mg GAE/g (fw) total phenols, and 219 mg QE/100G (fw) total flavonoids [37].

4. Health benefits of mulberry fruits

M. alba is a great composite of bioactive compounds, and due to their presence, this fruit has been used in Chinese medicine for treating fever, optical disorders, upper respiratory tract problems, high blood pressure, hypercholesterolemia, and joint problems [40]. In 1985 Ministry of Health of China had titled this fruit to be the first one to provide curative benefits to health. Korean and Japanese also use this fruit for the treatment of sore throat, liver and kidney disorders, and eye health issues [4].

4.1 Antioxidative activity

An antioxidant exhibits radical scavenging effects [41], the ability to transfer its electron to the radical to establish cell neutralization [42] and thus possess an antioxidizing nature [23] for the physiological processes reduction in oxidative stress caused due to pollutants, alcohol, medicine, illness, poor diet, toxins, and radiations is very much important because oxidative mutilation to DNA (deoxyribonucleic acid) proteins and other macromolecules becomes a resourceful reason for many diseases [30].

The phenolic content commands the antioxidative activity of the mulberry fruit, dependent on the cultivator, phenotype, amount of phytochemicals in fruit and age; as the fruit reaches the maturity age from unripened to ripened, the phenolic content increases. A vast variety and good quantity of polyphenols in mulberry fruit offer surplus benefits. The antioxidative capacity of mulberry fruit polysaccharides (MFP) is amazing, among four fractions of MFP, that is, MFP-1, MFP-2, MFP-3, and MFP-4, MFP-4 possess the best scavenging properties [4], in vitro studies mulberry fruit polysaccharide, that is, MFP3P-Se (selenide of polysaccharide) showed great radical scavenging activity as compared to MFP3P [43]. Four mulberry species, that is, alba, rubra, nigra, and laevigata, when tested for phenol and alkaloid content, showed a higher content of these compounds, that is, (880 ± 7.20)-1650 ± 12.25 mg/100 g fresh weight and (390 ± 3.22)-660 ± 5.25) mg/100 g fresh weight, respectively [44]. Mulberry fruit anthocyanins possess antioxidative properties verified by mulberry juice purification (MJP) and mulberry mark purification (MMP) [45]. Other than anthocyanins, mulberry fruits also contains resveratrol, a potent antioxidant; that reduces reactive oxygen species [46] and ultimately reduces DNA damage [47].

Due to its great composition, this fruit acts very strongly in radical striking areas, mainly skin, tissue, hair, and other areas of the body, thus preventing the oxidate stress of these radicals and ultimately reducing blemishes, age-related spots and promoting healthy body status [48] also gains the ability to prevent premature white hair formation [23], possess the ability to destruct oxidative substances that have a role in damaging red blood cells [49]. In context to mulberry fruit protective action, 70% ethanol extract comprised of 400 μg/mL showed a positive action against H_2O_2-induced oxidative injury of pancreatic MIN6N beta-cells [37]. In streptozotocin-induced rats, mulberry fruits boost up the antioxidative defense mechanism and lower down damaging oxidative substances [49]. The antioxidative potential of black mulberry fruits was assessed by preparing methanol extract and using α, α-diphenyl-β-picrylhydrazyln(DPPH). Results concluded the existence of high-level antioxidative activity of methanol extracts. In another experimental study, ethanol-ethanol water (1:1) and water extracts of black mulberry reported great antioxidative activity [50].

4.2 Anti-inflammatory activity

Mulberry fruits contain cyanidin-3-glucodise (C3 G) reported to provide anti-inflammatory properties, thus inhibiting inflammation and endothelial distinction. Researchers have also reported that black mulberries contain flavonoids providing anti-inflammatory response; work against proinflammatory cytokines in mice serum [51]. In female mice, tumor growth was inhibited by cyanidin-3-glucoside, ultimately leading to apoptosis [22].

4.3 Antibacterial activity

S. epidermidis and placacnes places are the bacterias responsible for acne; black mulberry fruit possesses antibacterial properties against these bacterias and has been reported to prevent bacterial acne [52]. *M. alba's* component; Morin, when investigated against bacterial activity, showed antibacterial properties against Streptococcus mutants [53]. Black mulberry possesses more antibacterial properties as compared to other mulberries, that is, studies revealed that flavonoids in black mulberry's possess great antibacterial properties against *Escherichia coli*, *Pseudomonas aeruginosa,* and *Staphylococcus aureus* than flavoinds present in other mulberry types [54].

4.4 Antiobesity activity

Obesity is the basic root cause of many chronological disorders, that is, diabetes mellitus, cardiovascular diseases, and cancer. Around the world risk of obesity and its related disorders is now increasing at a great speed which is now a point of concern. Mulberry fruit has been known to provide antiobesity properties. A study was conducted on male hamster rats to determine the effect of mulberry fruit extract on body weight; results concluded a decline in the body weight status [51].

4.5 Antitumor activity

To identify the role of natural compounds in possessing antitumor activities, a series of detailed research has been done for this purpose and came to the point that mulberry fruit offers virtuous antitumor activities through different pathways, an important biological quality [4]. Mulberry fruit extract (MFE) antitumor activity is characterized by the decline in nitric oxide production in lipopolysaccharides (LPS) produced RAW 264.7 cells, MFE in LPS also suppresses the expression of proinflammatory markers and cytokines [55]. Another study in mice also reported a positive role of MFE in suppressing acute colitis and tumorigenesis, and in S-phase cell cycle arrest, phenolic components of MFE provide a 70–80% decline in tumor growth [56]. MFE contains a good percentage of anthocyanins and, when fed to mice for the period of 7 weeks with an objective to investigate tumor growth, results concluded a decline in Atypical Glandular Cells (AGS) tumor xenograft growth [57]. Other than anthocyanins, resveratrol present in mulberry fruit has been reported to decrease the proliferative mechanisms of tumor growth in lymphoid, myeloid, breast, prostate, stomach, colon, pancreas, thyroid cancers, and ovarian and cervical cells [58].

4.6 Antihyperglycemic activity

Diabetes mellitus, characterized by high blood glucose levels due to defects in the secretion of insulin [51] and other metabolic issues and has now become the third lifespan-threatening disorder. It is also estimated that the prevalence of this disorder will constantly increase rather than decreasing; it is also expected that 300 million patients will be diabetic by 2025. Natural food items have been evaluated a lot for the determination of hypoglycemic effects to manage hyperglycemia [4]; mulberry fruit contains 1-deoxynojirimycin (DNJ), glycopeptides, hydrophobic flavonoids, and 2-arylbenzofuran; serves important role in high blood glucose management [30] provision of DNJ and alpha-glucosidase inhibitors; this fruit is very much fruitful against diabetes [30]. Four mulberry fruit polysaccharides have been isolated, that is, MFP-1, MFP-2, MFP-3 and MFP-4; MFP-3 and MFP-4 possess better inhibitory effects on α-amylase and α-glucosidase than MFP-1 and MFP-2 [43].

A study concluded that when diabetic rats were given mulberry fruit polysaccharides – MFP50 and MFP90 ethanol extract for the period of 7 weeks revealed 26.5% and 32.5% reduction in serum insulin levels and glucose levels, respectively. Another study conducted on diabetic rats showed that when they were given ethyl acetate-soluble extract of mulberry for the period of 2 weeks concluded a decrease in the fasting glucose and glycosylated serum protein levels. In another study, mulberry anthocyanin extract (50 and 125 mg/kg body weight per day) was taken to investigate its effects on blood sugar levels of diabetic mice, extract was given for the period of 8 weeks and showed an effective reduction in fasting blood glucose levels [37]. In streptozotocin-induced diabetic mice, mulberry fruit extract (MFE) has been reported to hamper alpha-glucosidase activity, suppress fasting blood glucose levels, and suppress glycosylated serum proteins [59]. A study conducted on two fractions of mulberry fruit polysaccharides, that is, MFP50 and MFP90, for the period of 7 weeks reported a decline in fasting glucose and serum insulin levels [60]. Another study on Ramulus mori polysaccharide obtained from white mulberry concluded the reduction in fasting glucose levels in diabetic rats [61]. Research conducted on diabetic C57BL/KsJ-db/db mice to determine the effect of MFE containing anthocyanins showed that it has the capability to improve insulin sensitivity, to decrease glucose 6-phosphatase, to decrease phosphoenolpyruvate carboxykinase ranks in the liver [4]. In streptozotocin-induced type 2 diabetic rats, polysaccharides present in mulberry fruit demonstrated a very chief role, that is, repaired damaged pancreatic tissue, lowered fasting glucose levels, decreased fasting insulin levels, and increased high-density lipoprotein cholesterol [60].

In the rats model, antidiabetic activity of mulberry anthocyanins has been observed, that is, a reduction in fasting blood glucose levels and serum insulin levels [62]. Increase in insulin sensitivity, activated protein kinase (AMPK), and glucose transporter type 4 (GLUT 4) and reduction of fasting hyperglycemia, postprandial blood glucose, and glycated serum proteins. Has been seen in chemically induced diabetic mice model given white mulberry fruit extract [63].

4.7 Cardioprotective activity

Resveratrol present in mulberry fruit provides protection against various heart-related diseases, decreasing oxidants levels by reducing HO and AKT levels and by increasing Trx-1 and vascular endothelial growth factor (VGEF) levels [47]. Anthocyanins in mulberry fruit possess antihyperlipidemic characteristics, that is, in rat models, a reduction in cholesterol, triglyceride, and adiponectin levels has been observed after administering this fruit [62].

Hyperlipidemia is defined by excessive cholesterol, that is, LDL (low-density lipoprotein) and VLDL (very low-density lipoprotein), and fatty materials in the blood leading to the development of atherosclerosis. A study conducted on human beings aged (30–60 years) with an objective to determine the effect of mulberry fruit consumption on lipid profiles concluded a positive association between lipid profile and mulberry fruit consumption, that is, decreasing total cholesterol and low-density lipoprotein, suggesting it for individuals with hypercholesterolemia [4]. Water extracts of mulberry fruit indicate lipid-lowering benefits by decreasing biosynthesis and increasing the levels of the LDLR gene. The presence of linoleic acid and dietary fiber in mulberry fruit provides a very positive action against hyperlipidemia; research conducted on rats provided them with 5% or 10% mulberry powder with a high-fat diet concluded a decrease in LDL, liver triglyceride, and total cholesterol and increase in HDL (high-density lipoprotein). Consumption of 45 g freeze-dried mulberries for 6 weeks by hyperlipidemic adults aged (30 to 60 years) quoted a very positive result against hyperlipidemia [37].

In another study conducted on Wistar rats to investigate the hypolipidemic activity of mulberry fruit freeze-dried powder, Wistar rats along with mulberry were given with high-fat diet, and the results concluded that their serum triglyceride, total cholesterol, serum low-density lipoprotein, liver triglycerides, and atherogenic profile were reduced, and levels of high-density lipoprotein cholesterol were increased [4]. In a study, 45 g of freeze-dried mulberry fruits were given to hypercholesterolemic individuals for the period of 6 weeks; a reduction in low-density lipoprotein (LDL) was observed [64]. Mulberry fruit extract also possesses a property against the development of atherosclerosis; in a study, rabbits induced with a high-fat diet, when administered with MFE, showed a reduction in atherosclerosis [65].

4.8 Nephroprotective activity

Mulberry fruit also possesses nephroprotective activity; butyl pyroglutamate, a bioactive compound obtained from mulberry fruit effective against cisplatin-induced nephrotoxicity [37]. A study conducted on rabbits to investigate the nephroprotective properties of *M. alba*; results concluded that after administering 200 mg/kg/d of *M. alba* ethanol extract for the period of 2 weeks subject's serum creatinine (4.02 ± 0.14, $p < 0.0001$), blood urea nitrogen (54.18 ± 2.60, $p < 0.0001$), and serum uric acid levels (2.34 ± 0.12, $p < 0.001$) were managed from the baseline results. Other than this, a decline in creatinine clearance and urinary volume was also reported [66].

4.9 Neuroprotective activity

Mulberry fruit consists of cyanidin-3-O-β-D-glucopyranoside; have a role in preventing neuronal cell damage and also prevents the cerebral ischemic damage caused by oxygen-glucose deprivation (OGD) in PC12 cells [67]. Researchers observed mulberry fruit ripening stages and found that ripened mulberry fruit possesses great neuroprotective potential due to the occurrence of high concentrations of phenolic compounds [68]. It also possesses antistress action by inhibiting plasma peroxide levels [37]. Its fruit extract, when tested in animal models with vascular dementia against memory impairment and brain damage, was observed that it is a very potent natural cognitive enhancer and neuroprotectant [69]. Mulberry fruit extract also contains Oxyreservatrol, a natural hydroxystilbene that provides a neuroprotective effect against neurodegenerative diseases caused due to oxidative damage such as Alzheimer's disease, characterized by cellular damage, high lipid peroxidation and an increase in iron and aluminum levels [70]. Cyanidine-3-glucoside (C3G) fraction from *M. alba* possess neuroprotective potential, and when analyzed for oxygen deprivation and glutamate-induced cell death in rats, results concluded it possesses a protective potential against oxygen-depressed cellular loss not on glutamate-induced cell death [71].

Omega 3 and 6 are essential fatty acids important for the development of healthy cell membranes, and for boosting neural functions by the formation of hormones, eicosanoids, thromboxanes, leukotrienes, and prostaglandins [72]. The human body cannot synthesize these essential fatty acids, and thus exogenous source is required; mulberry fruit contains these fatty acids, which then provide strength to brain functioning and protection against cholesterol neurotoxicity [73]. In the mammalian central nervous system, dopaminergic neurons present in the midbrain are the chief source of dopamine (DA); loss of dopaminergic neurons leads to a neurological disorder, that is, Parkinson's disease (PD) [74]. Mulberry fruit extract has been found protective for dopaminergic neurons against neurotoxicity, thus protecting against Parkinson's disease, ultimately ameliorating chronic disorders [75]. In addition, mulberry fruit extract has been found supportive in rat models facing memory-related problems and hippocampal damage, that is, MFE improved their memory and also heightened the concentrations of neurons and cholinergic activity [76].

4.10 Gastroprotective activity

Mulberry fruit contains iron, riboflavin, vitamin C, vitamin K, potassium, phosphorous, calcium, dietary fiber, and phytonutrients, that is, zeaxanthin, resveratrol, anthocyanins, lutein, and polyphenolic compounds. Dietary fiber plays an essential role in the prevention of constipation, bloating, and cramping by adding bulk to the stool and by speeding up the movement of food through the digestive tract [77]. In experimental rats, gastric mucosal injury characterized by an observable fall in leucocyte infiltration to the submucosal layer was significantly reduced by *M. alba* extract [78]. *M. nigra* fruit's oral administration of a high dose (300 mg/kg) is very potent against ethanol-induced acute gastric ulcers in female rats [79]. It has been reported in mice this fruit effectively prevented constipation; along with this, it also increased the concentration of acetic, propionic, butyric, isovaleric, lactobacillus, and bifidobacterium and decreased the occurrence of *Helicobacter* and prevotellaceae in feces [80].

4.11 Cancer-protective activity

Many medicinal plants exhibit protective properties against bacteria, virus, inflammation, and cancer, thus providing immune strengthening and antioxidative properties. Anthocyanins obtained from *M. alba* have shown inhibition of metastatic A549 human lung cancerous cells [81]. Phenolic compounds possess anticancer properties in in vitro hepatoma cells by arresting cell cycle at G2-M phase and by stopping topo-isomerase II activity [38]. International Agency for Research on Cancer (IARC) specialized agency of cancer by the World Health Organization (WHO), stated that 10 million deaths were caused due to cancer worldwide by 2020; Asia ranked in the top (58.3%), followed by Europe (19.6%) due to resistance to drugs and their side effects [82]. In animal tumor studies, phenols present in mulberry depicted anticarcinogenic mechanisms in different cancerous cells, that is, working as an antioxidant, an antiproliferative, and an antiangiogenic, thus leading to the induction of apoptosis. Anthocyanins such as cyanidin and pelargonidin present in *M. nigra* possess the ability to decrease the capability of cancerous cells and also possess the ability to hinder tumor growth [22]. Resveratrol present in mulberry fruit suppresses the activity of transcription factor 4EBP1 and also decreases the proliferation of cancerous cells in breast cells [51]. *M. nigra*; a well-known anthocyanin-rich fruit, possesses the quality of inhibiting the formation of stomach cancer caused by *Helicobacter pylori*; a prominent factor in causing stomach cancer [39].

4.12 Hepatoprotective activity

Accumulation of lipids in liver cells is the leading cause of the development of nonalcoholic fatty liver. In mice, hepatic steatosis caused by a hyperlipid diet has been improved by ethanolic extract of *M. nigra* fruits, that is, reduction in serum levels of ALT and AST, reduction in lipid droplets of liver cells, reduction in triglyceride and cholesterol levels of liver cells and reduction in fatty acid and cholesterol biosynthesis [83]. A study was conducted on three mulberry fruit polysaccharides, that is, MFP-I, MFP-II, and MFP-III, by adding 30%, 60%, and 90% ethanol; the mixture was then given to evaluate the efficacy of MFP against palmitic acid (PA) induced liver lipotoxicity, the study concluded MFP-II significantly decreased PA-induced liver lipotoxicity at 0.1 and 0.2 mg/ml and also enhanced catalase and glutathione peroxidase activities; protecting against PA-induced liver lipotoxicity [84].

5. Conclusion

This chapter focused on the fruit named mulberry, which is very known and commonly used around the world for its extraordinary health benefits. Many different types of mulberry plants exist, but this chapter focused on its three major types, that is, *M. alba* (white mulberry), *M. nigra* (black mulberry) and *M. rubra* (Red mulberry). These all are very well known due to great nutritional composition as composition shows the presence of phytonutrients & bioactive compounds, that is, anthocyanins, flavanols, flavonols, phenolic compounds, aromatic compounds, volatile compounds, resveratrol, cyanidin-3-O-β-D-glucopyranoside, cyanidin-3-glucodise (C3 G), butyl pyroglutamate, and melatonin. The presence of these phytonutrients and bioactive compounds made this fruit responsible for offering multiple health-promoting and protecting activities, that is, antioxidation, anti-inflmmation, antitumor, antibacterial, antiobesity, antihyperglycemic, cardioprotective, nephroprotective, neuroprotective, hepatoprotective, cancer-protective, and gastroprotective.

Appendices

AGS	Atypical Glandular Cells
AMPK	Activated Protein Kinase
C3 G	Cyanidin-3-glucodise
DNA	Deoxyribonucleic Acid
GLUT 4	Glucose transporter type 4
HDL	High-Density Lipoprotein
IARC	International Agency for Research on Cancer
LDL	Low-Density Lipoprotein
LPS	Lipopolysaccharides
MFP	Mulberry Fruit Polysaccharides
MJP	Mulberry Juice Purification

MMP	Mulberry Mark Purification
PD	Parkinson's disease
VLDL	Very Low-Density Lipoprotein
VGEF	Vascular Endothelial Growth Factor
WHO	World Health Organization
OGD	Oxygen Glucose Deprivation

Author details

Zoha Sohail
FMH Institute of Allied Health Sciences, Pakistan

*Address all correspondence to: zoha1sohail@gmail.com

References

[1] Hashemi S, Khadivi A. Morphological and pomological characteristics of white mulberry (*Morus alba* L.) accessions. Scientia Horticulturae. 2020;**259**:108827

[2] Andreu MG, Friedman MH, McKenzie M, Quintana HV. Morus rubra, red mulberry. EDIS. 2010;**2010**(4):1-2

[3] Ozrenk K, Gazioglu SR, Erdinc C, Guleryuz M, Aykanat A. Molecular characterization of mulberry germplasm from Eastern Anatolia. African Journal of Biotechnology. 2010;**9**(1):2-5

[4] Yuan Q, Zhao L. The Mulberry (*Morus alba L.*) fruit a review of characteristic components and health benefits. Journal of Agricultural and Food Chemistry. 2017;**65**(48):10383-10394

[5] Gungor N, Sengul M. Antioxidant activity, total phenolic content and selected physicochemical properties of white mulberry (*Morus alba L.*) fruits. International Journal of Food Properties. 2008;**11**(1):44-52

[6] Türkyılmaz M, Özkan M. Mulberry Juice. Handbook of Functional Beverages and Human Health: CRC Press; 2016. pp. 423-432

[7] Singhal BK, Khan MA, Dhar A, Baqual FM, Bindroo BB. Approaches to industrial exploitation of mulberry (Mulberry sp.) fruits. Journal of Fruit and Ornamental Plant Research. 2010;**18**(18):83-99

[8] Doymaz İ. Pretreatment effect on sun drying of mulberry fruits (*Morus alba L.*). Journal of Food Engineering. 2004;**65**(2):205-209

[9] Krishna H, Singh D, Singh RS, Kumar L, Sharma BD, Saroj PL. || Morphological and antioxidant characteristics of mulberry (Morus spp.) genotypes. Journal of the Saudi Society of Agricultural Sciences. 2020;**19**(2):136-145

[10] Erturk Y, Gecer M, editors. Economy of Berries. Proceedings of 4th National Berry Symposium; 2012

[11] Gunes M, Cekic C. Some chemical and physical properties of fruits of different mulberry species commonly grown in Anatolia, Turkey. Asian Journal of Chemistry. 2004;**16**(3):1849-1855

[12] Rodriguez-Mateos A, Vauzour D, Krueger CG, Shanmuganayagam D, Reed J, Calani L, et al. Bioavailability, bioactivity and impact on health of dietary flavonoids and related compounds: An update. Archives of Toxicology. 2014;**88**(10):1803-1853

[13] Pawlowska AM, Oleszek W, Braca A. Quali-quantitative analyses of flavonoids of Morus nigra L. and *Morus alba L.*(Moraceae) fruits. Journal of Agricultural and Food Chemistry. 2008;**56**(9):3377-3380

[14] Devi B, Sharma N, Kumar D, Jeet K. *Morus alba* Linn: A phytopharmacological review. International Journal of Pharmacy and Pharmaceutical Sciences. 2013;**5**(2):14-18

[15] Elmacı Y, Altuğ T. Flavour evaluation of three black mulberry (Morus nigra) cultivars using GC/MS, chemical and sensory data. Journal of the Science of Food and Agriculture. 2002;**82**(6):632-635

[16] Mehta M, Kumar A. Nutrient composition, phytochemical profile and antioxidant properties of Morus

nigra: A Review. International Journal of Innovative Science and Research Technology. 2021;**6**(2):424-432

[17] Okatan V. Phenolic compounds and phytochemicals in fruits of black mulberry (Morus nigra L.) genotypes from the Aegean region in Turkey. Folia Horticulturae. 2018;**30**(1):93

[18] Naderi GA, Asgary S, Sarraf-Zadegan N, Oroojy H, Afshin-Nia F. Antioxidant activity of three extracts of Morus nigra. Phytotherapy Research: An International Journal Devoted to Pharmacological and Toxicological Evaluation of Natural Product Derivatives. 2004;**18**(5):365-369

[19] Okatan V, Polat M, Aşkin MA. Some physico-chemical characteristics of black mulberry (Morus nigra L.) in bitlis. Scientific Papers-Series B. Horticulture. 2016;**60**:27-30

[20] Sengül M, Ertugay MF, Sengül M. Rheological, physical and chemical characteristics of mulberry pekmez. Food Control. 2005;**16**(1):73-76

[21] Parida S, Rayaguru K, Panigrahi J. Mulberry cultivation and its phytochemical benefits: A review. Journal of Natural Remedies. 2020;**21**(5):33-48

[22] Memete AR, Timar AV, Vuscan AN, Venter AC, Vicas SI. Phytochemical composition of different botanical parts of Morus species, health benefits and application in Food Industry. Plants. 2022;**11**(2):152

[23] Rohela GK, Shukla P, Kumar R, Chowdhury SR. Mulberry (Morus spp.): An ideal plant for sustainable development. Trees, Forests and People. 2020;**2**:100011

[24] Jiang Y, Nie W-J. Chemical properties in fruits of mulberry species from the Xinjiang province of China. Food Chemistry. 2015;**174**:460-466

[25] Sánchez-Salcedo EM, Mena P, García-Viguera C, Hernández F, Martínez JJ. (Poly) phenolic compounds and antioxidant activity of white (*Morus alba*) and black (Morus nigra) mulberry leaves: Their potential for new products rich in phytochemicals. Journal of Functional Foods. 2015;**18**:1039-1046

[26] Tiwari R, Rana C. Plant secondary metabolites: A review. International Journal of Engineering Research and General Science. 2015;**3**(5):661-670

[27] Kim SB, Chang BY, Jo YH, Lee SH, Han S-B, Hwang BY, et al. Macrophage activating activity of pyrrole alkaloids from *Morus alba* fruits. Journal of Ethnopharmacology. 2013;**145**(1):393-396

[28] Zhu J, Wang L, Xiao Z, Niu Y. Characterization of the key aroma compounds in mulberry fruits by application of gas chromatography–olfactometry (GC-O), odor activity value (OAV), gas chromatography-mass spectrometry (GC–MS) and flame photometric detection (FPD). Food Chemistry. 2018;**245**:775-785

[29] Sánchez-Salcedo EM, Mena P, García-Viguera C, Martínez JJ, Hernández F. Phytochemical evaluation of white (*Morus alba* L.) and black (Morus nigra L.) mulberry fruits, a starting point for the assessment of their beneficial properties. Journal of Functional Foods. 2015;**12**:399-408

[30] Sharma S, Bisla G. Nutritional and health benefits of mulberry (Morus nigra L.). International Journal of Food and Nutritional Sciences. 2019;**8**(1):20

[31] Rao AV, Snyder DM. Raspberries and human health: A review. Journal

[47] De La Lastra CA, Villegas I. Resveratrol as an antioxidant and pro-oxidant agent: Mechanisms and clinical implications. Biochemical Society Transactions. 2007;**35**(5):1156-1160

[48] Prasath CH, Balasubramanian A, Radhakrishnan S. Mulberry fruit-Nutritional and health benefits. 2019;**6**:15-16

[49] Hong J-H, Ahn J-M, Park S-W, Rhee S-J. The effects of mulberry fruit on the antioxidative defense systems and oxidative stress in the erythrocytes of streptozotocin-induced diabetic rats. Nutritional Sciences. 2004;**7**(3):127-132

[50] Kostiæ D, Dimitrijeviæ D, Mitiæ S, Mitiæ M, Stojanoviæ G, Živanoviæ A. Phenolic content and antioxidant activities of fruit extracts of Morus nigra L (Moraceae) from Southeast Serbia. Tropical Journal of Pharmaceutical Research. 2013;**12**(1):105-110

[51] Ramappa VK, Srivastava D, Singh P, Kumar U, Kumar D, Gosipatala SB, et al. Mulberries: A promising fruit for phytochemicals, nutraceuticals, and biological activities. International Journal of Fruit Science. 2020;**20**(sup3):S1254-S1S79

[52] Budiman A, Aulifa DL, Kusuma ASW, Sulastri A. Antibacterial and antioxidant activity of black mulberry (Morus nigra L.) extract for acne treatment. Pharmacognosy Journal. 2017;**9**(5):611-614

[53] Yang J-Y, Lee H-S. Evaluation of antioxidant and antibacterial activities of morin isolated from mulberry fruits (*Morus alba* L.). Journal of Korean Society for Applied Biological Chemistry. 2012;**55**:485-489

[54] Chen H, Yu W, Chen G, Meng S, Xiang Z, He N. Antinociceptive and antibacterial properties of anthocyanins and flavonols from fruits of black and non-black mulberries. Molecules. 2017;**23**(1):4

[55] Qian Z, Wu Z, Huang L, Qiu H, Wang L, Li L, et al. Mulberry fruit prevents LPS-induced NF-κB/pERK/MAPK signals in macrophages and suppresses acute colitis and colorectal tumorigenesis in mice. Scientific Reports. 2015;**5**(1):1-13

[56] Alam AK, Hossain AS, Khan MA, Kabir SR, Reza MA, Rahman MM, et al. The antioxidative fraction of white mulberry induces apoptosis through regulation of p53 and NFκB in EAC cells. PLoS One. 2016;**11**(12):e0167536

[57] Huang H-P, Chang Y-C, Wu C-H, Hung C-N, Wang C-J. Anthocyanin-rich Mulberry extract inhibit the gastric cancer cell growth in vitro and xenograft mice by inducing signals of p38/p53 and c-jun. Food Chemistry. 2011;**129**(4):1703-1709

[58] Aggarwal BB, Bhardwaj A, Aggarwal RS, Seeram NP, Shishodia S, Takada Y. Role of resveratrol in prevention and therapy of cancer: Preclinical and clinical studies. Anticancer Research. 2004;**24**(5A):2783-2840

[59] Wang Y, Xiang L, Wang C, Tang C, He X. Antidiabetic and antioxidant effects and phytochemicals of mulberry fruit (*Morus alba* L.) polyphenol enhanced extract. PLoS One. 2013;**8**(7):e71144

[60] Jiao Y, Wang X, Jiang X, Kong F, Wang S, Yan C. Antidiabetic effects of *Morus alba* fruit polysaccharides on high-fat diet-and streptozotocin-induced type 2 diabetes in rats. Journal of Ethnopharmacology. 2017;**199**:119-127

[61] Guo C, Li R, Zheng N, Xu L, Liang T, He Q. Anti-diabetic effect of ramulus

of Agricultural and Food Chemistry. 2010;**58**(7):3871-3883

[32] Jung S, Lee M-S, Choi A-J, Kim C-T, Kim Y. Anti-inflammatory effects of high hydrostatic pressure extract of mulberry (*Morus alba*) fruit on LPS-stimulated RAW264. 7 cells. Molecules. 2019;**24**(7)

[33] Łochyńska M. Energy and nutritional properties of the white mulberry (Morus alba L.). Journal of Agricultural Science and Technology. 2015;**5**:2-13

[34] Zaikov GE, Pudel D, Spychalski G. Renewable resources and biotechnology for material applications. Renewable Resources and Biotechnology for Material Applications. 2011:1-410

[35] Ercisli S, Orhan E. Chemical composition of white (*Morus alba*), red (Morus rubra) and black (Morus nigra) mulberry fruits. Food Chemistry. 2007;**103**(4):1380-1384

[36] Jan B, Parveen R, Zahiruddin S, Khan MU, Mohapatra S, Ahmad S. Nutritional constituents of mulberry and their potential applications in food and pharmaceuticals: A review. Saudi Journal of Biological Sciences. 2021;**28**(7):3909-3921

[37] Dhiman S, Kumar V, Mehta C, Gat Y, Kaur S. Bioactive compounds, health benefits and utilisation of Morus spp.– A comprehensive review. The Journal of Horticultural Science and Biotechnology. 2020;**95**(1):8-18

[38] Naowaratwattana W, De-Eknamkul W, De Mejia EG. Phenolic-containing organic extracts of mulberry (*Morus alba L.*) leaves inhibit HepG2 hepatoma cells through G2/M phase arrest, induction of apoptosis, and inhibition of topoisomerase IIα activity. Journal of Medicinal Food. 2010;**13**(5):1045-1056

[39] Naeem MY. Medicinal potentials and health benefits of black mulberry. Eurasian Journal of Food Science and Technology. 2020;**4**(1):1-5

[40] Katsube T, Imawaka N, Kawano Y, Yamazaki Y, Shiwaku K, Yamane Y. Antioxidant flavonol glycosides in mulberry (*Morus alba L.*) leaves isolated based on LDL antioxidant activity. Food Chemistry. 2006;**97**(1):25-31

[41] Chung Y-C, Chen S-J, Hsu C-K, Chang C-T, Chou S-T. Studies on the antioxidative activity of Graptopetalum paraguayense E. Walther. Food Chemistry. 2005;**91**(3):419-424

[42] Olszowy M. What is responsible for antioxidant properties of polyphenolic compounds from plants? Plant Physiology and Biochemistry. 2019;**144**:135-143

[43] Chen C, You L-J, Abbasi AM, Fu X, Liu RH, Li C. Characterization of polysaccharide fractions in mulberry fruit and assessment of their antioxidant and hypoglycemic activities in vitro. Food & Function. 2016;7(1):530-539

[44] Imran M, Khan H, Shah M, Khan R, Khan F. Chemical composition and antioxidant activity of certain Morus species. Journal of Zhejiang University. Science. B. 2010;**11**:973-980

[45] Jiang D-Q, Guo Y, Xu D-H, Huang Y-S, Yuan K, Lv Z-Q. Antioxidant and anti-fatigue effects of anthocyanins of mulberry juice purification (MJP) and mulberry marc purification (MMP) from different varieties mulberry fruit in China. Food and Chemical Toxicology. 2013;**59**:1-7

[46] Pandey KB, Rizvi SI. Anti-oxidative action of resveratrol: Implications for human health. Arabian Journal of Chemistry. 2011;**4**(3):293-298

models. The British Journal of Nutrition. 2010;**104**(1):8-16

[76] Kaewkaen P, Tong-Un T, Wattanathorn J, Muchimapura S, Kaewrueng W, Wongcharoenwanakit S. Mulberry fruit extract protects against memory impairment and hippocampal damage in animal model of vascular dementia. Evidence-based Complementary and Alternative Medicine. 2012;**2012**

[77] Kadam RA, Dhumal ND, Khyade VB. The Mulberry, *Morus alba* (L.): The medicinal herbal source for human health. International Journal of Current Microbiology and Applied Sciences. 2019;**8**(4):2941-2964

[78] Abdulla MA, Ali HM, Ahmed KA-A, Noor SM, Ismail S. Evaluation of the anti-ulcer activities of *Morus alba* extracts in experimentally-induced gastric ulcer in rats. Biomedical Research. 2009;**20**(1):35-39

[79] Nesello LAN, Beleza MLML, Mariot M, Mariano LNB, de Souza P, Campos A, et al. Gastroprotective value of berries: Evidences from methanolic extracts of Morus nigra and Rubus niveus fruits. Gastroenterology Research and Practice. 2017;**2017**

[80] Hu T-G, Wen P, Fu H-Z, Lin G-Y, Liao S-T, Zou Y-X. Protective effect of mulberry (Morus atropurpurea) fruit against diphenoxylate-induced constipation in mice through the modulation of gut microbiota. Food & Function. 2019;**10**(3):1513-1528

[81] Sharma P, Sharma A, Thakur J, Murali S, Bali K. Mulberry as a life savior - A review. Journal of Pharmacognosy and Phytochemistry. 2020;**9**(2):2445-2451

[82] Wild C, Weiderpass E, Stewart BW. World cancer report: Cancer research for cancer prevention. International Agency for Research on cancer Cancer. 2020:246-247

[83] Song H, Lai J, Tang Q, Zheng X. Mulberry ethanol extract attenuates hepatic steatosis and insulin resistance in high-fat diet–fed mice. Nutrition Research. 2016;**36**(7):710-718

[84] Hu D, Bao T, Lu Y, Su H, Ke H, Chen W. Polysaccharide from mulberry fruit (*Morus alba* L.) protects against palmitic-acid-induced hepatocyte lipotoxicity by activating the Nrf2/ARE signaling pathway. Journal of Agricultural and Food Chemistry. 2019;**68**(46):13016-13024

mori polysaccharides, isolated from *Morus alba* L., on STZ-diabetic mice through blocking inflammatory response and attenuating oxidative stress. International Immunopharmacology. 2013;**16**(1):93-99

[62] Yan F, Dai G, Zheng X. Mulberry anthocyanin extract ameliorates insulin resistance by regulating PI3K/AKT pathway in HepG2 cells and db/db mice. The Journal of Nutritional Biochemistry. 2016;**36**:68-80

[63] Choi KH, Lee HA, Park MH, Han J-S. Mulberry (*Morus alba* L.) fruit extract containing anthocyanins improves glycemic control and insulin sensitivity via activation of AMP-activated protein kinase in diabetic C57BL/Ksj-db/db mice. Journal of Medicinal Food 2016;19(8):737-745.

[64] Sirikanchanarod A, Bumrungpert A, Kaewruang W, Senawong T, Pavadhgul P. The effect of mulberry fruits consumption on lipid profiles in hypercholesterolemic subjects: A randomized controlled trial. Journal of Pharmacy and Nutrition Sciences. 2016;**60**:7-14

[65] Chen C-C, Liu L-K, Hsu J-D, Huang H-P, Yang M-Y, Wang C-J. Mulberry extract inhibits the development of atherosclerosis in cholesterol-fed rabbits. Food Chemistry. 2005;**91**(4):601-607

[66] Ullah N, Khan MA, Khan S, Ahmad H, Asif AH, Khan T. Nephro-protective potential of *Morus alba*, a prospective experimental study on animal models. Pharmaceutical Biology. 2016;**54**(3):530-535

[67] Kang TH, Hur JY, Kim HB, Ryu JH, Kim SY. Neuroprotective effects of the cyanidin-3-O-β-d-glucopyranoside isolated from mulberry fruit against cerebral ischemia. Neuroscience Letters. 2006;**391**(3):122-126

[68] Yang J, Liu X, Zhang X, Jin Q, Li J. Phenolic profiles, antioxidant activities, and neuroprotective properties of mulberry (Morus atropurpurea Roxb.) fruit extracts from different ripening stages. Journal of Food Science. 2016;**81**(10):C2439-C2C46

[69] Bohnen NI, Albin RL. The cholinergic system and Parkinson disease. Behavioural Brain Research. 2011;**221**(2):564-573

[70] Rodrigues EL, Marcelino G, Silva GT, Figueiredo PS, Garcez WS, Corsino J, et al. Nutraceutical and medicinal potential of the Morus species in metabolic dysfunctions. International Journal of Molecular Sciences. 2019;**20**(2):301

[71] Bhuiyan MIH, Kim H-B, Kim S-Y, Cho K-O. The neuroprotective potential of cyanidin-3-glucoside fraction extracted from mulberry following oxygen-glucose deprivation. The Korean Journal of Physiology and Pharmacology. 2011;**15**(6):353-361

[72] Feng Y, Liu M, Ouyang Y, Zhao X, Ju Y, Fang Y. Comparative study of aromatic compounds in fruit wines from raspberry, strawberry, and mulberry in central Shaanxi area. Food & Nutrition Research. 2015;**59**(1):29290

[73] El-Sayyad HI. Cholesterol overload impairing cerebellar function: The promise of natural products. Nutrition. 2015;**31**(5):621-630

[74] Chinta SJ, Andersen JK. Dopaminergic neurons. The International Journal of Biochemistry & Cell Biology. 2005;**37**(5):942-946

[75] Kim HG, Ju MS, Shim JS, Kim MC, Lee S-H, Huh Y, et al. Mulberry fruit protects dopaminergic neurons in toxin-induced Parkinson's disease

Chapter 8

Phytochemical Study and Biological Activities of Endemic Species of *Vaccinium* from French Polynesia

Sarah Benayad, Stéphanie Soulet and Taivini Teai

Abstract

French Polynesia is a natural laboratory with over 53% of endemism in its vascular terrestrial flora. The flora remains little studied from a phytochemical point of view. In order to overcome this lack of knowledge, we were interested in an endemic taxon from the specie *Vaccinium cereum*, also named as "opu opu." Four varieties of *V. cereum* are present in French Polynesia: var. *adenandrum* (Decne) F.Br, var. *cereum* (L.f.) G. Forst, var. *pubiflorum* Skottsb and var. *raiateense* M.L. Grant. All four showed comparable antioxidant and antiradical activities and inhibitory activity against α-glucosidase. Leaves of *V. cereum* var. *cereum* carry out a bio-guided phytochemical study. Among the isolated compounds, NMR has characterized ursolic acid, oleanolic acid, chlorogenic acid, astragalin, and isoquercitrin, as the major active compounds. The results revealed that this taxon represents a real health benefit and might have promising proprieties to regulate blood sugar.

Keywords: *Vaccinium cereum*, phytochemical studies, biological activities, endemism, French Polynesia

1. Introduction

French Polynesia is a French overseas collectivity composed of 118 islands spread over an area as large as continental Europe. Its geographical situation, its isolation, and its tropical climate give French Polynesia a rich and original terrestrial flora.

The native vascular flora of French Polynesia has about 870 species, of which 460 are endemic, or 54% of this flora. Some species are endemic to an archipelago, an island, or even a valley or a mountain. Conserving this biodiversity, as well as finding the right balance between protecting and benefiting from this resource has been a widely accepted concept since the adoption of the Convention on Biological Diversity (CBD) in Rio in 1992.

Diabetes represents one of the largest health problems in the world, affecting more and more people each year, with an estimated total burden of death due to high blood sugar in 2020 of 4 million [1, 2].

In 2017, according to figures from the IDF (International Diabetes Federation), 425 million people had diabetes. IDF estimates that the number of people with diabetes will increase by 48% worldwide by 2045. These rates of increase may reach 82% in Southeast Asia and 156% in Africa [3].

In French Polynesia, 10% of the population has diabetes. This disease remains the second most common chronic noncommunicable disease after hypertension. The number of patients diagnosed in 10 years (from 2001 to 2012) has doubled and continues to increase. This disease raises real social, political, and economic issues for the future of French Polynesia: it cost nearly 30 million euros in health care expenses, added to various patient complications.

Diabetes is a health priority in French Polynesia [4]. In addition to the diabetes campaigns conducted by the French Polynesian Health Department for more than 15 years, the search for new treatments remains important, both locally and globally.

Numerous plant-based preparations are described in traditional medicine for the prevention of diabetes but have not been the subject of real treatments due to a lack of laboratory studies. In the last decades, several works on the mode of action and the targets of these preparations could revive the interest in the use of plants in the treatment and prophylaxis of type two diabetes, T2DM.

Table 1 describes some plants or plant preparations used to treat T2DM. This list is not exhaustive; it cites plants used in traditional medicine worldwide and studied for their antidiabetic effect.

Many plants are used for their hypoglycemic effect. In general, the preparations combine several plants and can be administered in different forms: dry extracts (capsules, herbal teas), essential oils, or mother tinctures. In addition, in the context of antidiabetic treatment, the use of antioxidants in supplementation helps prevent the appearance of complications related to oxidative stress [5–11].

In this context, the first research program allowed to perform a screening on different biological targets from several extracts of terrestrial and marine organisms targeting Alzheimer's disease, melanoma, and diabetes. Thus, pharmacological screening on enzymatic targets (acetylcholinesterase, α-glucosidase, and tyrosinase) subject 117 marine extracts and 228 extracts from Polynesian terrestrial flora (from 122 species).

This work has allowed highlighted the strong inhibitory activity of the ethanolic extract of *Vaccinium cereum* var. *cereum* leaves toward α-glucosidase, an enzyme involved in the regulation of blood sugar and the control of T2DM.

2. *Vaccinium* sp.

The genus *Vaccinium* (bilberries, blueberries, and cranberries) is a member of the Ericaceae family, which includes more than 340 species in 87 genera.

In 1933, there were 160 species belonging to the genus *Vaccinium* distributed between the northern hemisphere and the mountains of tropical regions.

In 2014, Northern Hemisphere described 450 species. Many of them are cultivated for their berries as they are often edible.

The antidiabetic properties of plants of the genus *Vaccinium* were first cited in 1892. The infusion of bilberry leaves would decrease glycosuria [12].

Research on the use of the aqueous extract of *Vaccinium* as an antidiabetic agent mobilized the scientific world until 1936, with suspicion of the presence of toxic hydroquinones. In 1996, *Vaccinium myrtillus* demonstrated the absence of hydroquinone, which re-launched research on the antidiabetic properties of *Vaccinium* extracts' [12].

Plant	Preparation	Plant part	Use and therapeutic target
Allium cepa	Raw, infusion, mother tincture	Bulbs	Hypoglycemic, antioxidant
Allium sativum	Raw, freeze-dried powder, alcoholic tincture	Bulbs	Hypoglycemic
Anacardium sp.	Ethanolic extract	Barks	Hypoglycemic *in vivo*
Artocarpus altilis	Methanolic extract	Bark and fruits	Inhibition of α-glucosidase and α-amylase [5]
Artocarpus heterophyllus	Methanolic extract	Sheets	Inhibition of α-glucosidase and α-amylase [5]
Cinnamomum zeylanicum	Methanolic extract	Sheets	Inhibition of α-glucosidase and α-amylase [5]
Eucalyptus globulus	Powder	Sheets	Hypoglycemic
Ficus benghalensis	Decoction	Barks	Hypoglycemic *in vivo* [6]
Moringa oleifera	Ethanolic extract	Sheets	Hypoglycemic *in vivo*
Piper betle	Methanolic extract	Sheets	Inhibition of α-glucosidase and α-amylase [5]
Phyllanthus amarus	Ethanolic extract and decoction	Sheets	Reduction of blood glucose [7]
Prosopis africana	Ethanolic extract	Barks	Hypoglycemic *in vivo*
Punica granatum	Methanolic extract	Flowers	Inhibition of α-glucosidase [8]
Ramulus mori	Decoction	Branches	Inhibition of α-glucosidase [9]
Rosa sinensis	Infusion	Flowers	Hypoglycemic
Sclerocarya birrea	Ethanolic extract	Roots	Hypoglycemic *in vivo*
Syzygium cumini	Aqueous extract	Branches	Hypoglycemic, α-amylase inhibitor [10]
Vaccinium angustifolium	Alcoholic extract	Sheets	Insulin-like and glitazone-like. Promotes the proliferation of pancreatic cells
Vaccinium myrtillus	Alcoholic extract	Sheets	Hypoglycemic

Table 1.
Use of herbs in the treatment of type 2 diabetes.

Table 2 summarizes some recent research on the antidiabetic properties of organic extracts and molecular families isolated from different *Vaccinium* species.

2.1 *Vaccinium cereum* (L.f.) G.Forst

In French Polynesia, two endemic species, whose berries are edible, are present [22]: *Vaccinium cereum* (L.f.) G.Forst and *Vaccinium rapae* Skottsb.

V. rapae being an endangered species listed on the red list of threatened species in France, the study was conducted only on the species *V. cereum* (**Figure 1**).

Several vernacular names are used to refer to *Vaccinium cereum* as "opu opu" in Tahiti, "heua" or "heuki" in Nuku Hiva and "puatoatoa" in Fatu Hiva [22]. It is endemic to Polynesia. Present in the Society archipelago, the Marquesas

Species	Plant part	Extract/family of molecules studied	Mode of action
Vaccinium angustifolium	Roots	Ethanolic extract	Transport of glucose to cells [13]
	Barks	Ethanolic extract	Transport of glucose to cells, stimulates insulin secretion [13]
	Sheets	Ethanolic extract	Transport of glucose to cells, stimulates insulin secretion [13]
	Fruits	Anthocyanins	Proliferation of β-pancreatic cells [13] Stimulates insulin secretion [14]
Vaccinium arctostaphylos	Fruits	Hydroxycinnamic acid, phenolic compounds anthocyanins	Inhibits α-glucosidase and stimulates insulin secretion [15]
	Sheets	Flavonoids anthocyanins	Inhibition of α-amylase Stimulates insulin secretion [16]
Vaccinium ashei	Sheets	Glucosidic flavonol, substituted phenylpropanoid-3-ols, anthocyanins, proanthocyanidins	Inhibition of α-glucosidase [17, 18]
	Fruits	Aqueous acetone extract (50%)	Inhibition of α-glucosidase [19]
Vaccinium bracteatum	Sheets	Polysaccharides	Stimulates insulin secretion [19]
Vaccinium corymbosum var. jersey	Fruits	Aqueous acetone extract (50%)	Inhibition of α-glucosidase [19]
	Flowers	Hydroxycinnamic acid, glucoside flavonol, phenylpropanoid-substituted catechin	Inhibition of α-glucosidase [19]
Vaccinium myrtillus	Sheets	Aqueous ethanolic extract (40%)	Transport of glucose to cells [20]
Vaccinium vitis-idaeae	Fruits	Glucosidic flavonols	Transport of glucose to cells [21]

Table 2.
Organic extracts and hypoglycemic molecules isolated from the genus Vaccinium.

archipelago, and the Cook Islands, *V. cereum* grows at altitudes above 500 m, notably up to the summit of Mount Aorai (2064 m), one of the highest peaks of the island of Tahiti. Observed also at lower altitudes, the shrubs are smaller.

It is a shrub of 2 to 2.5 m in height. The wood is very hard and quite dense. The leaves on the old branches are large, leathery, and elliptical (5 × 3.5 cm), while they are thinner, smaller, and generally obovate (average 2.3 × 1.6 cm) on the flowering stems. They are persistent for 3–4 years, shiny above, paler below.

The flowers are permanent throughout the year. They are axillary, solitary, on a stalk 5 to 8 mm, and about 1 cm long. The corolla is white, or with a pink to red tip [22] and a cinnamon-vanilla perfume.

The fruits, edible but very little consumed, are red, dark purple to black, of approximately 1.5 cm in diameter. The pubescence of the flowers is very variable. On this last criterion, Skottsberg determined a certain number of varieties or taxons.

In French Polynesia, four taxa of *Vaccinium cereum* are recognized (**Figure 2**):

- *Vaccinium cereum* var. *adenandrum* (Decne.) F.Br., endemic to the Marquesas Islands;

a)

b)

Figure 1.
a) Flower buds and b) fruit of Vaccinium cereum *var.* cereum *(Photo credit S. Benayad).*

- *Vaccinium cereum* var. *cereum*, (L.f.) G.Forst., endemic to the Society Archipelago;
- *Vaccinium cereum* var. *pubiflorum* Skottsb, endemic to the Society Archipelago;
- *Vaccinium cereum* var. *raiateense* M.L.Grant, endemic to the island of Raiatea (Society Archipelago).

In order to enhance *Vaccinium cereum*: A comparative study of the biological activity of the different taxa found in French Polynesia (*V.c.* var. *cereum*, *V.c.* var. *adenandrum*, *V.c.* var. *pubiflorum, V.c.* var. *raiateense*) has been carried out *V.c.* var. *raiateense* being an endangered taxon listed on the red list of threatened species in France only a few leaves were collected for a comparative analysis of their biological activities with the other three taxa.

The biological targets chosen were diabetes (α-glucosidase inhibition test) and oxidative stress (DPPH, $ABTS^{+\cdot}$, ORAC, and Folin-Ciocalteu tests). We also chose to focus, when available, on different plant parts: roots, branches, stems, leaves, flowers, and fruits.

Finally, a phytochemical study of the leaves of the taxon *V.c.* var. *cereum* was conducted to identify the secondary metabolites present in the ethanolic extract.

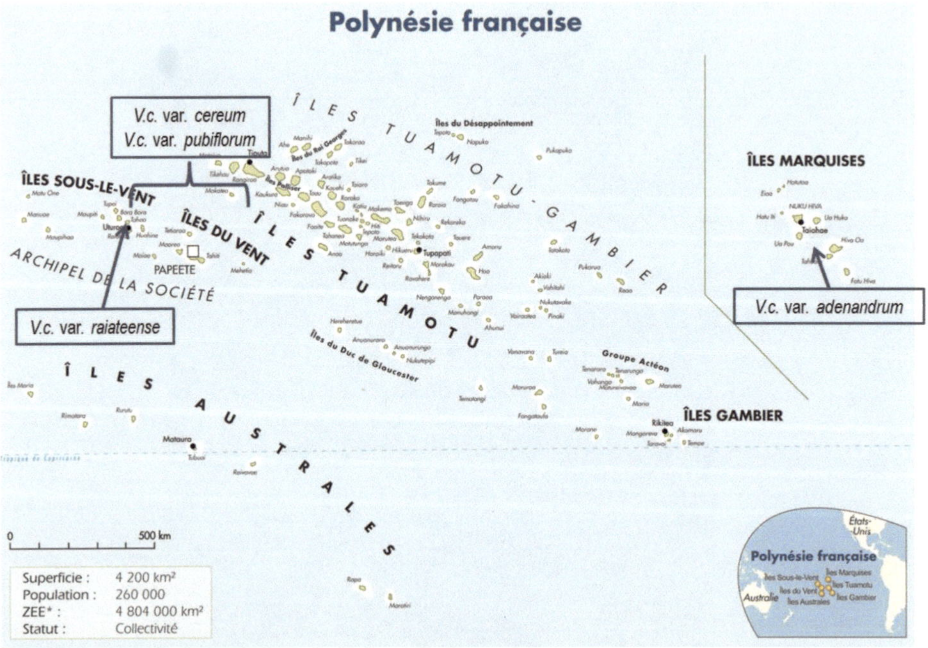

Figure 2.
Distribution map of Vaccinium cereum *taxa in French Polynesia.*

2.2 Screening of biological activities of different taxa of *Vaccinium cereum*

Sampling was performed between December 2014 and August 2016, with 53 samples collected:

- Seven samples of *Vaccinium cereum* var. *adenandrum* were collected on the island of Fatu Hiva (Marquesas archipelago);
- Thirty-three samples of *Vaccinium cereum* var. *cereum* were collected from Mount Marau and Mount Aorai on the island of Tahiti (Society Archipelago);
- Nine samples of *Vaccinium cereum* var. *pubiflorum* were collected from Mount Orohena on the island of Tahiti (Society Archipelago) and;
- Four samples of *Vaccinium cereum* var. *raiateense* were taken on the Temehani-Rahi plateau on the island of Raiatea (Society archipelago).

Samples from the different plant parts of these taxa were extracted by maceration in 96° ethanol for eight hours. After evaporation of the ethanol, the crude extracts obtained were tested on the selected biological targets.

Table 3 details the obtained results. The yields of the crude extracts vary from 3–26% based on the initial dry matter mass. The mass yield of ethanolic extracts from leaves is the highest, with an average of 20%, those from fruits vary from 4–25% yield and those from roots vary from 3–17%.

The results show that all the extracts, without distinction of variety or part of plant, present a strong inhibitory potential of – glucosidase with IC_{50} ranging between 0.5 and

Taxon	Plant part	Sample name	Location	Date	Yield extract (%)	Inhibition of α-glucosidase	
						α-glucosidase inhibition at 10 μg/mL [1]	IC_{50} of α-glucosidase in μg/mL [2]
V.c. var. *adenandrum*	Branches	Vca B	Fatu Hiva	December 14	9	98.8 ± 0.6	2.1 ± 0.7
	Sheets	Vca Fe1	Fatu Hiva	December 14	16.3	96.9 ± 0.5	5.3 ± 0.4
		Vca Fe2	Fatu Hiva	July 16	16.2	74.9 ± 7.1	4.5 ± 1.0
		Vca Fe3	Fatu Hiva	July 16	19.3	95.7 ± 1.9	0.9 ± 0.2
	Green fruits	Vca Gf	Fatu Hiva	December 14	4.4	100.0 ± 0.1	3.1 ± 0.9
	Roots	Vca R	Fatu Hiva	December 14	17.4	99.6 ± 0.2	0.7 ± 0.1
	Stems	Vca S	Fatu Hiva	December 14	14.8	99.7 ± 0.2	5.4 ± 0.2
V.c. var. cereum	Branches	Vcc B2	Tahiti – Aorai	February 15	4.1	100.0 ± 0.2	2.5 ± 0.4
		Vcc B3	Tahiti – Aorai	June 15	8.4	99.0 ± 0.5	1.4 ± 0.5
		Vcc B4	Tahiti – Aorai	June 15	5.3	98.9 ± 0.7	1.2 ± 0.4
		Vcc B5	Tahiti – Aorai	June 15	8.8	99.1 ± 0.6	1.0 ± 0.1
	Sheets	Vcc Fe1	Tahiti – Aorai	November 14	17.5	97.3 ± 0.2	6.1 ± 0.5
		Vcc Fe3	Tahiti – Aorai	June 15	15.1	97.3 ± 0.3	3.8 ± 0.5
		Vcc Fe4	Tahiti – Aorai	June 15	19.1	96.8 ± 0.5	5.0 ± 0.6
		Vcc Fe5	Tahiti – Aorai	June 15	17.4	96.8 ± 0.5	3.9 ± 0.6
		Vcc JFe6	Tahiti – Aorai	June 15	18.6	95.5 ± 0.6	7.2 ± 1.0
		Vcc Fe7	Tahiti – Marau	July 15	11.1	96.5 ± 0.4	7.2 ± 0.2
		Vcc Fe8	Tahiti – Marau	July 15	13.5	98.9 ± 0.3	4.0 ± 0.5
		Vcc Fe9	Tahiti – Marau	July 15	16.7	97.3 ± 1.2	4.1 ± 1.0
		Vcc Fe10	Tahiti – Aorai	July 16	20.1	95.7 ± 1.2	1.3 ± 0.2
		Vcc Fe11	Tahiti – Aorai	July 16	28.9	97.8 ± 0.2	0.7 ± 0.1
	Flowers	Vcc Fl1	Tahiti – Aorai	November 14	12.4	55.8 ± 5.9	9.8 ± 0.2
		Vcc Fl3	Tahiti – Aorai	June 15	16.8	98.6 ± 0.3	8.1 ± 0.3
		Vcc Fl6	Tahiti – Aorai	June 15	11	98.7 ± 0.4	7.6 ± 0.4
	Fruits	Vcc Fr1	Tahiti – Aorai	November 14	22.3	76.8 ± 1.3	8.8 ± 0.1
		Vcc Fr2	Tahiti – Aorai	February 15	5	58.4 ± 2.6	9.6 ± 0.1
		Vcc Fr3	Tahiti – Aorai	June 15	25.5	98.2 ± 0.4	7.8 ± 0.2
		Vcc Fr6	Tahiti – Aorai	June 15	21.5	86.6 ± 1.6	8.6 ± 01
		Vcc Fr9	Tahiti – Marau	July 15	10.3	4.7 ± 0.8	ND
	Green fruits	Vcc Frv9	Tahiti – Marau	July 15	3.2	36.3 ± 2.2	ND
	Roots	Vcc R2	Tahiti – Aorai	February 15	5.3	99.0 ± 0.4	0.8 ± 0.1
		Vcc R7	Tahiti – Marau	July 15	3	98.6 ± 0.5	1.1 ± 0.4
	Stems	Vcc T1	Tahiti – Aorai	November 14	11.5	98.8 ± 1.2	2.3 ± 0.3

Taxon	Plant part	Sample name	Location	Date	Yield extract (%)	Inhibition of α-glucosidase	
						α-glucosidase inhibition at 10 μg/mL [1]	IC_{50} of α-glucosidase in μg/mL [2]
V.c. var. *cereum*	Stems	Vcc T3	Tahiti – Aorai	June 15	16.6	98.1 ± 0.9	3.7 ± 0.5
		Vcc T4	Tahiti – Aorai	June 15	11.6	98.3 ± 0.7	1.3 ± 0.3
		Vcc T5	Tahiti – Aorai	June 15	14.2	97.9 ± 1.4	3.0 ± 0.7
		Vcc T6	Tahiti – Aorai	June 15	16.7	99.5 ± 0.3	5.6 ± 0.4
		Vcc T7	Tahiti – Marau	July 15	4.7	100.0 ± 0.1	1.0 ± 0.2
		Vcc T8	Tahiti – Marau	July 15	4.4	100.0 ± 0.4	1.1 ± 0.2
		Vcc T9	Tahiti – Marau	July 15	2.8	98.9 ± 0.6	2.9 ± 0.4
V.c. var. *pubiflorum*	Branches	Vcp B	Tahiti – Orohena	July 15	7.7	99.8 ± 0.1	1.7 ± 0.9
	Sheets	Vcp Fe	Tahiti – Orohena	July 15	26.2	98.8 ± 0.2	2.9 ± 0.7
		Vcp Fe1	Tahiti – Aorai	August 16	26.3	100.0 ± 0.2	1.3 ± 0.1
		Vcp Fe2	Tahiti – Aorai	August 16	25.1	99.4 ± 0.2	0.8 ± 0.1
		Vcp Fe3	Tahiti – Aorai	August 16	22.2	99.6 ± 0.1	1.1 ± 0.2
	Flowers	Vcp Fl	Tahiti – Orohena	July 15	14.7	97.6 ± 0.4	7.9 ± 0.3
	Fruits	Vcp Fr	Tahiti – Orohena	July 15	17	96.8 ± 0.2	7.9 ± 0.3
	Roots	Vcp R	Tahiti – Orohena	July 15	7.3	99.7 ± 0.6	0.7 ± 0.1
	Stems	Vcp T	Tahiti – Orohena	July 15	9	99.4 ± 0.3	2.4 ± 0.3
V.c. var. *raiateense*	Sheets	Vcr Fe1	Raiatea	August 16	22.3	98.7 ± 0.1	0.6 ± 0.1
		Vcr Fe2	Raiatea	August 16	35	98.1 ± 0.2	1.2 ± 0.1
		Vcr Fe3	Raiatea	August 16	18.1	97.8 ± 0.2	0.5 ± 0.0
		Vcr Fe4	Raiatea	August 16	18.4	98.4 ± 0.2	1.4 ± 0.2

[1]: α-glucosidase inhibition of acarbose at 300 μg/mL = 56.7 ± 4.21%. [2]: IC_{50} of acarbose = 270.5 ± 3.73 μg/mL; ND: not determined.

Table 3.
Inhibitory potential of α-glucosidase of ethanolic extracts of Vaccinium cereum *taxa.*

9.6 μg/mL. These activities are very significant compared to that of the reference product used here and marketed in therapeutics, acarbose, which has an IC_{50} of 270.5 μg/mL.

Figure 3 represents the average IC50 on α-glucosidase obtained for the different plant parts of *Vaccinium cereum* taxa. The crude extracts of roots, branches, and stems have the highest activity toward α-glucosidase inhibition with mean IC_{50} of 0.8 ± 0.2 μg/mL; 1.7 ± 0.6 μg/mL; and 2.7 ± 1.3 μg/mL, respectively. Leaves have intermediate activity with an average IC_{50} of 4.9 ± 1.5 μg/mL.

The activity of the crude flower and fruit extracts was weaker, with average IC_{50} of 8.4 ± 1.0 μg/mL and 8.2 ± 0.9 μg/mL, respectively. These activities are still interesting compared to the inhibitory potential of acarbose (IC_{50} of 270.5 ± 3.73 μg/mL).

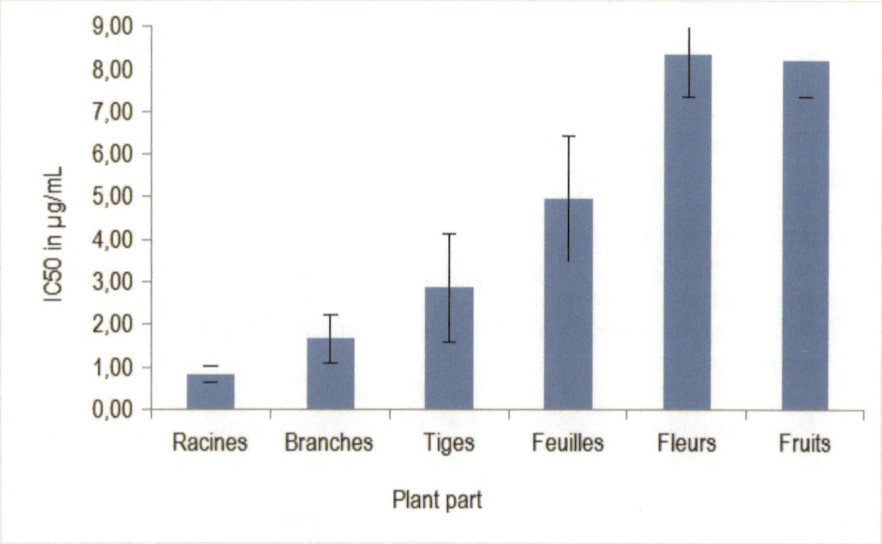

Figure 3.
Graphical representation of the average IC_{50} of extracts from different plant parts of Vaccinium cereum *taxa on α-glucosidase. N = 4 for roots (Racines), 6 for branches (branches), 10 for stems (Tiges), 21 for leaves (Feuilles), 4 for flowers (Fleurs) and 8 for fruits (fruits).*

Figure 4 represents the average IC_{50} of leaf extracts for the four *Vaccinium cereum* taxa present in French Polynesia.

Thus, the average IC_{50} of the leaves of *pubiflorum* and *raiateense* varieties appear to be the lowest with 1.5 µg/mL and 0.9 µg/mL, respectively. In contrast, the IC_{50} of the varieties *adenandrum* and *cereum* vary from 0.9 to 5.3 µg/mL, and from 0.7 to 7.2 µg/mL, respectively, and show less inhibition compared to the other two varieties.

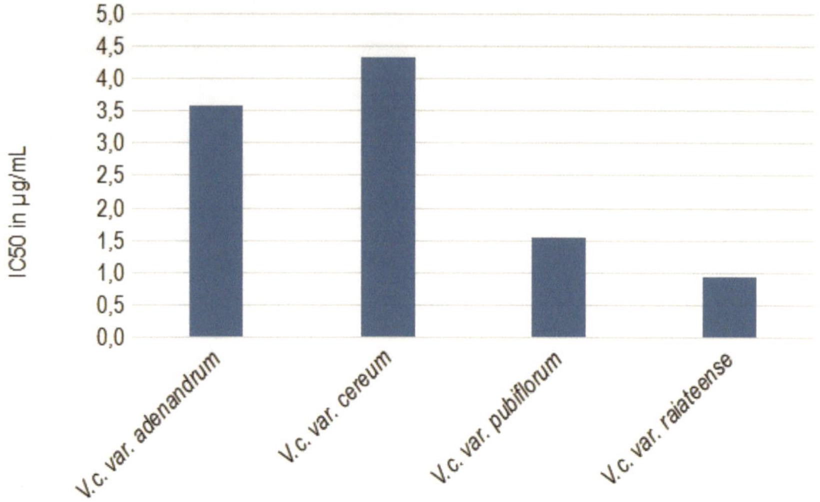

Figure 4.
Graphical representation of the average IC_{50} of leaf extracts of different Vaccinium cereum taxa on α-glucosidase. N = 3 for V.c. *var.* adenandrum, *10 for* V.c. *var.* cereum, *4 for* V.c. *var.* pubiflorum *and 4 for* V.c. *var.* raiateense.

The study of the antioxidant and antiradical potential was also carried out on the 53 samples of *V. cereum*. For this purpose, different complementary methods were used, namely:

- The radical recombination test to evaluate the antiradical potential of the extracts (two radicals tested: $ABTS^{+\cdot}$ and DPPH);
- The determination of total polyphenols by the Folin-Ciocalteu method allows to estimate the quantity of polyphenols present in the extracts. The correlation of this assay with the radical recombination test (DPPH and $ABTS^{+\cdot}$) allows to know if the observed antiradical activity is associated with the presence of polyphenols in the extracts, and;
- The measurement of the ORAC index, allowing to evaluate the antioxidant capacity of the extracts.

The results obtained are presented in the following figures.

The EC_{50} of the recombination of ethanolic extracts with DDPH vary from 71.4 μg/mL to 5.5 μg/mL for the most active extracts. The highest activity is observed for root and branch extracts with average EC_{50} of 5.8 μg/mL and 8.4 μg/mL, respectively.

The same distribution of recombination activities is obtained with $ABTS^{+\cdot}$, with root extracts again being the lowest EC_{50} (3.6 μg/mL and 4.3 μg/mL, respectively) and therefore showing the highest potential (**Figure 5**).

The EC_{50} on recombination with DPPH of leaf extracts from the four *V. cereum* taxa varied only slightly from 18.6 to 24.8 μg/mL. In contrast, the average EC_{50} on $ABTS^{+\cdot}$ of leaf extracts of the varieties *raiateense* and *pubiflorum*, 4.8 and 6.9 μg/mL, is lower than that of leaf extracts of the varieties *adenandrum* and *cereum*, 12.0 and 14.6 μg/mL, respectively (**Figure 6**).

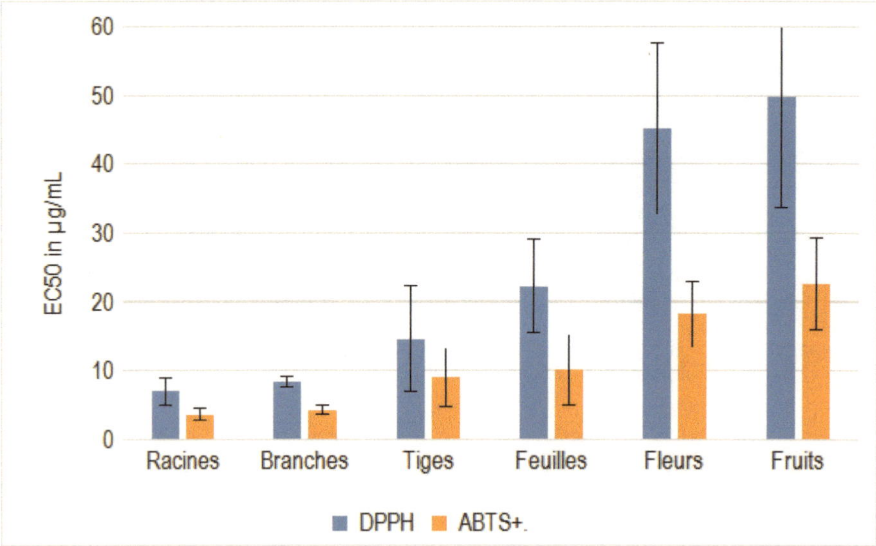

Figure 5.
Histogram of the average EC_{50} of the extracts of the different plant parts of Vaccinium cereum *taxa on the recombination of DPPH and $ABTS^{+\cdot}$ Radicals. N = 4 for roots (Racines), 6 for branches (branches), 10 for stems (Tiges), 21 for leaves (Feuilles), 4 for flowers (Fleurs) and 8 for fruits (fruits).*

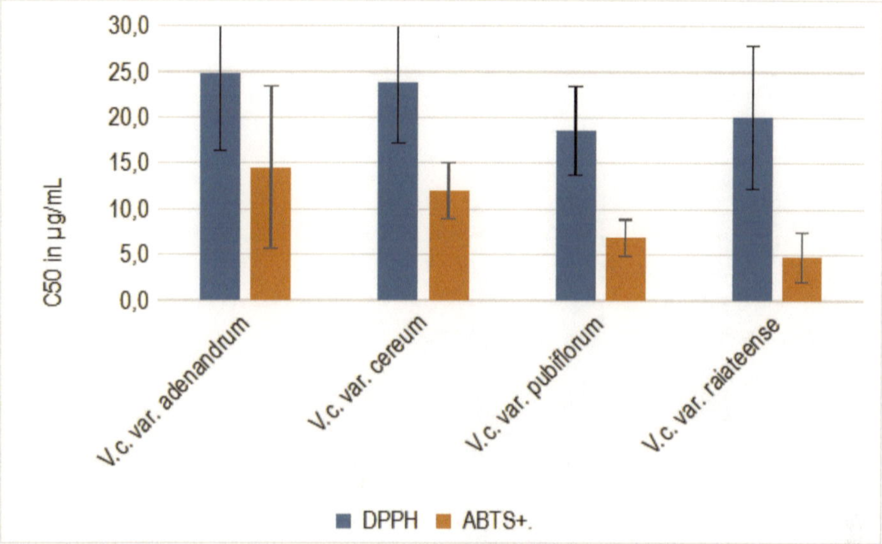

Figure 6.
Histogram of the average EC_{50} of leaf extracts of the four Vaccinium cereum *taxa on the recombination of DPPH and $ABTS^{+\cdot}$ Radicals. N = 3 for* V.c. *var.* adenandrum, *10 for* V.c. *var.* cereum, *4 for* V.c. *var.* pubiflorum *and 4 for* V.c. *var.* raiateense.

For the Folin-Ciocalteu assay, all samples were assayed and the results obtained are expressed as catechin equivalent and gallic acid equivalent per 100 g of extract (expressed as percentage).

Thus, all samples contain polyphenols, but there are large differences with values ranging from 2.4% in catechin equivalent for the fruit extract of *V.c.* var. *cereum* (Vcc Fr2) to 89% in catechin equivalent for the root extract of *V.c.* var. *pubiflorum* (VcpR). Roots and stems are the plant parts containing the most polyphenols while fruits show a low concentration (**Figure 7**). However, the latter is described to be rich in anthocyanosides. The Folin-Ciocalteu method, although widely used, does not allow an accurate determination of polyphenols because the reagent is sensitive to any type of reductants (such as oses, proteins, etc.) [23].

The leaves of V.*c.* var. *raiateense* and *pubiflorum* are the richest in polyphenols with an average percentage of 42% and 37% in catechin equivalent. The least rich in polyphenols are the varieties *cereum* and *adenandrum*, with average percentages of 28% and 24% in catechin equivalent (**Figure 8**).

The ORAC antioxidant potential of all extracts was measured, the results obtained are expressed in µmol trolox equivalent/mg of extract. All samples have a good antioxidant capacity with values ranging from 11.6 µmol trolox equivalent/mg for the most active extract to 1.2 µmol trolox equivalent/mg extract. These values confirm the strong antioxidant potential of *V. cereum* compared to the vitamin C index whose value is equal to 9.4 µmol trolox equivalent/mg.

All the extracts from the different parts of the plants seem to have the same antioxidant power (**Figure 9**), without any distinction being made on the ORAC index, especially for the leaves of the four taxa (**Figure 10**).

Vaccinium cereum is a wild shrub not cultivated in French Polynesia. The different taxa of *Vaccinium cereum* showed a very strong biological potential on the different biological targets chosen. The objective being to valorize this plant as a food

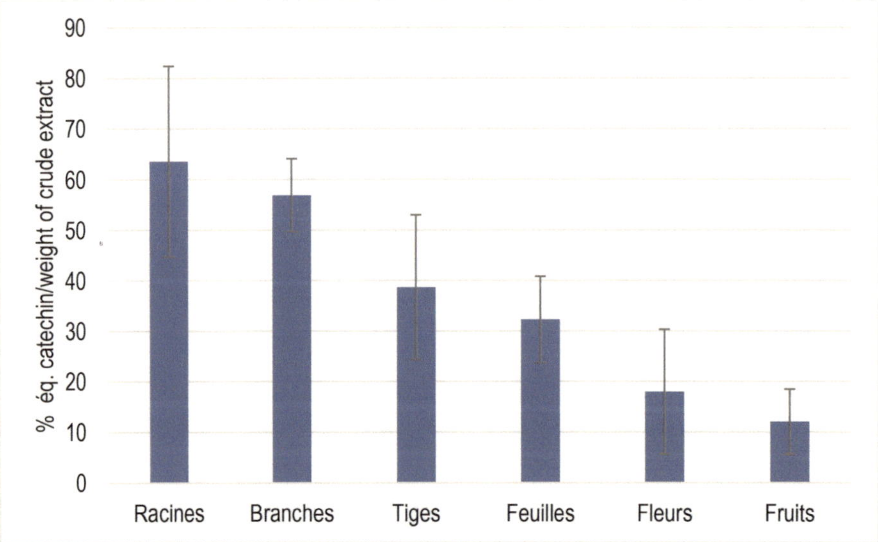

Figure 7.
Average percentage of total polyphenols in the crude extracts of the four Vaccinium cereum *taxa. N = 4 for roots (Racines), 6 for branches (Branches), 10 for stems (Tiges), 21 for leaves (Feuilles), 4 for flowers (Fleurs) and 8 for fruits (Fruits).*

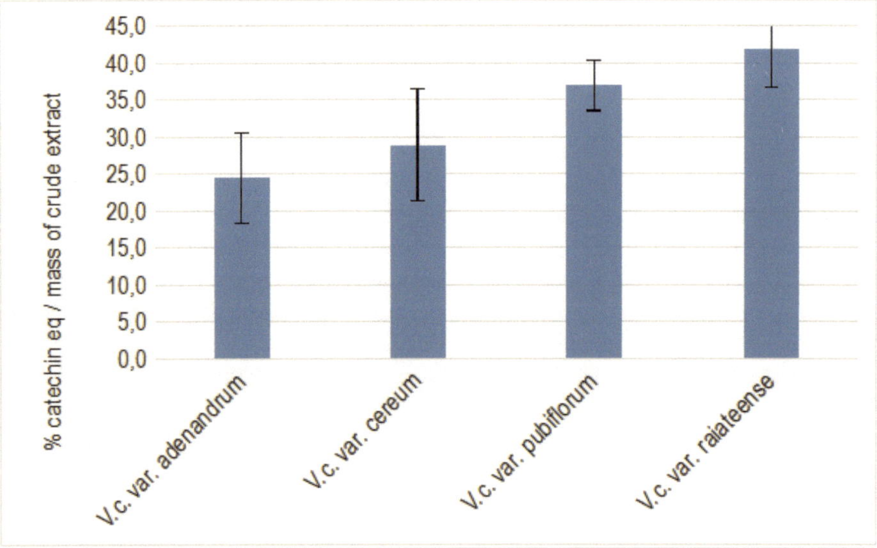

Figure 8.
Average percentage of total polyphenols in crude extracts of Vaccinium cereum *leaves. N = 3 for* V.c. *var.* adenandrum, *10 for* V.c. *var.* cereum, *4 for* V.c. *var.* pubiflorum *and 4 for* V.c. *var.* raiateense.

supplement to control glycemia, a phytochemical study was necessary to identify the secondary metabolites.

Among the taxa studied, *Vaccinium cereum* var. *cereum*, the most available and accessible, was chosen for this phytochemical study. Moreover, the evaluation of the inhibitory activity of α – glucosidase showed that all plant parts had a strong

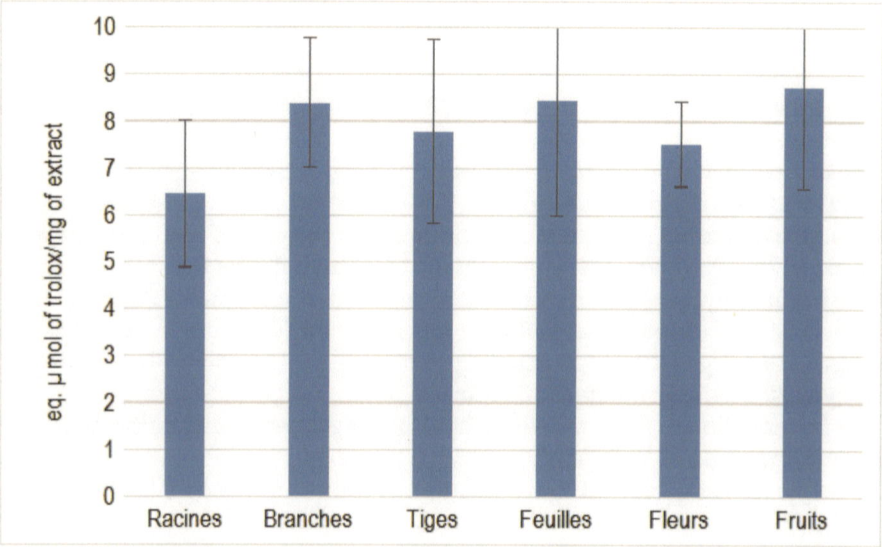

Figure 9.
Average ORAC index of extracts from the different plant parts of the four Vaccinium cereum *taxa. N = 4 for roots (Racines), 6 for branches (Branches), 10 for stems (Tiges), 21 for leaves (Feuilles), 4 for flowers (Fleurs) and 8 for fruits (Fruits).*

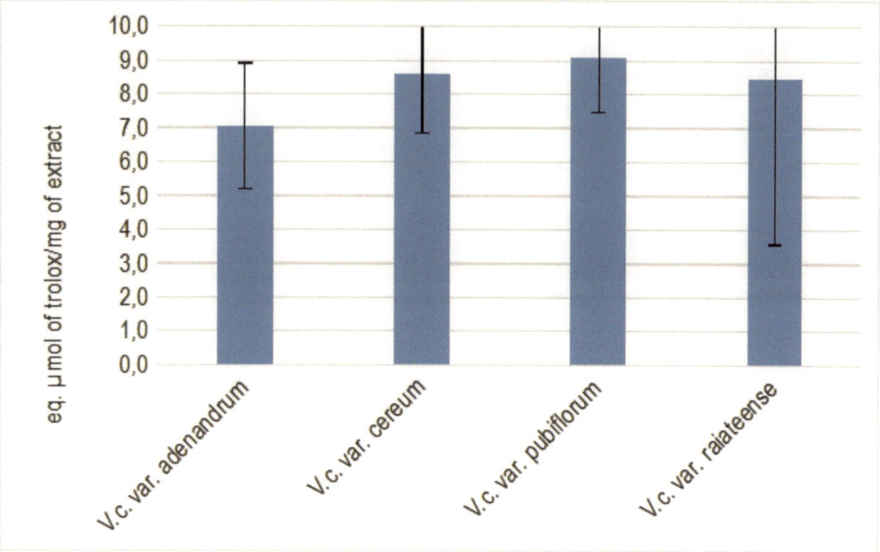

Figure 10.
Average ORAC index of leaf extracts of the four Vaccinium cereum *taxa. N = 3 for* V.c. *var.* adenandrum, *10 for* V.c. *var.* cereum, *4 for* V.c. *var.* pubiflorum *and 4 for* V.c. *var.* raiateense.

inhibitory potential with low IC_{50}, lower than acarbose, the positive control. However, analysis of the data revealed a higher activity for roots and branches.

However, we chose to work on the leaves of *Vaccinium cereum* var. *cereum* because they show high activity, interesting extraction yield, and better availability (renewable resource).

2.3 Study and identification of secondary metabolites isolated from leaves of *Vaccinium cereum* var. *cereum*

The taxon was collected in November 2014 on Mount Aorai, Tahiti Island, Society Archipelago.

The dry powder of *V. c.* var. *cereum* leaves was macerated in 96% ethanol at room temperature. The resulting crude extract (SB02) was taken up in distilled water and then fractionated using liquid-liquid partitioning by solvents of increasing polarity: cyclohexane, dichloromethane, ethyl acetate, and butanol. The four organic phases and the residual aqueous phase were then concentrated to dryness (**Figure 11**).

Screening for α-glucosidase inhibition was performed. The crude extract as well as the phases from liquid/liquid partitioning, showed activities above 80% inhibition of the enzyme at a concentration of 10 μg/mL.

The fractionation of the dichloromethane phase allowed the identification of two major compounds (Vc1 and Vc2).

The ethyl acetate phase led to the isolation of four compounds (Vc3 to Vc6).

2.3.1 Molecules isolated from the dichloromethane phase

2.3.1.1 Vc1 compound

A white amorphous powder was obtained, consisting of 95% of compound Vc1. The proton NMR of this compound is complex in the 0.5 to 2.5 ppm region and is characteristic of highly saturated compounds such as triterpenes with structures close to ursolic acid (**Figure 12** and **Table 4**).

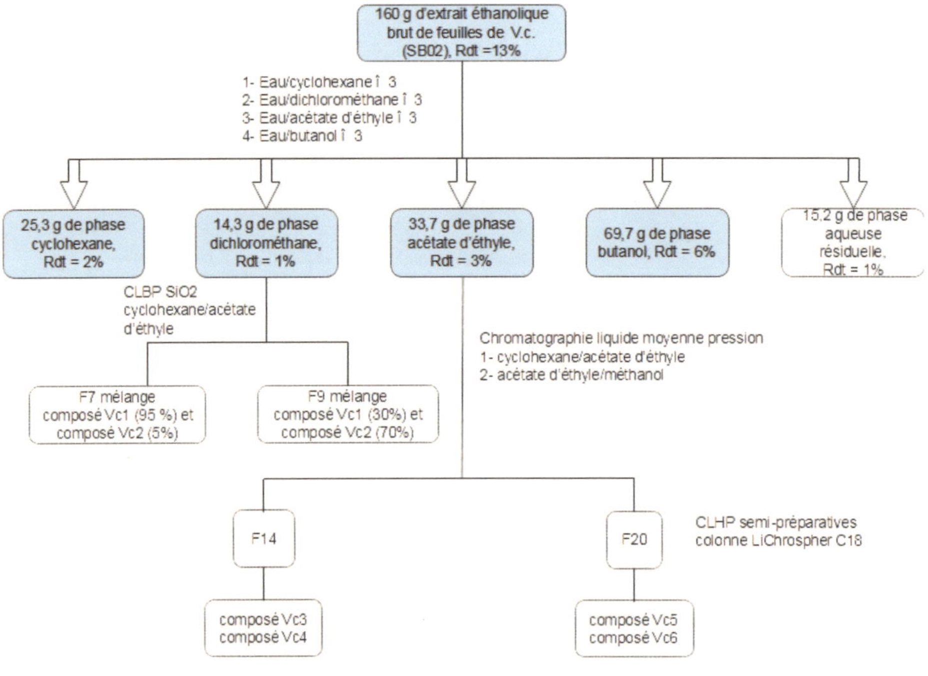

Figure 11.
Fractionation of the extract of leaves of Vaccinium cereum *var.* cereum.

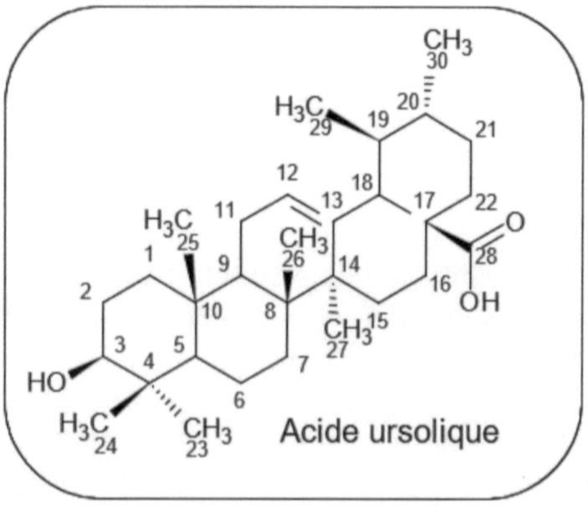

Figure 12.
Molecular structure of ursolic acid (Acide ursolique) extract from leaves of Vaccinium cereum *var.* cereum.

Compound Vc1: Ursolic acid		
	Experimental data	**Literature [24]**
Position	δ1H (ppm, CD3OD)	δ1H (ppm, CDCl3)
3	3.16 (1H, dd, J = 11.5 and 4.3 Hz)	3.21 (1H, dd, J = 4.4 Hz and 10.2 Hz)
12	5.23 (1H, t, J = 3.9 Hz)	5.28 (1H, t, J = 3.6 Hz)
18	2.20 (1H, d, J = 11.5 Hz)	2.18 (1H, d, J = 11.7 Hz)
23	1.29 (3H, s)	1.25 (3H, s)
24	0.96 (3H, s)	0.98 (3H, s)
25	0.78 (3H, s)	0.77 (3H, s)
26	0.98 (3H, s)	1.08 (3H, s)
27	1.12 (3H, s)	1.14 (3H, s)
29	0.88(3H, d, J = 6.5 Hz)	0.93 (3H, d, J = 6.5 Hz)
30	0.96(3H, d, J = 4.5 Hz)	0.91 (3H, d, J = 5.9 Hz)

Table 4.
Comparison of NMR data 1H of compound Vc1 and ursolic acid.

A triplet at 5.23 ppm is characteristic of the trisubstituted ethylenic group (H12) and coupled to two protons. The doublet at 2.20 ppm can be attributed to the proton carried by a tertiary carbon in α of a double bond and coupled to a single proton. Finally, in the area from 0.7 to 2.5 ppm five singlets and two doublets can be distinguished, proving the presence of 7 methyl groups. This skeleton would be that of ursolic acid (the spectra being recorded in MeOD, the proton of the acid function could not be distinguished).

The purity of the compound was measured at 95% by integration of the methyl signal at position 26. The IC_{50} of the mixture with 95% of this compound was determined for α-glucosidase and was 10.9 ± 0.9 μg/mL, proving significant activity compared with the control acarbose (270.5 μg/mL).

Ursolic acid is present in many plants and found in other species of the genus *Vaccinium* such as *V. myrtillus*, *Vaccinium macrocarpon*, *V. oxycocos*, *Vaccinium angustifolium* and *Vaccinium vitis-idaea* [25–28].

Ursolic acid is known to have several pharmacological uses including anticancer of breast [29], liver [30], colon [31], prostate [26], blood [32], antioxidant [33], anti-inflammatory [34], antifilarial [35], hypolipidemic and liver-protective [36].

2.3.2 Vc2 compound

Compound Vc2 is a white amorphous powder. The proton NMR of this compound as for compound 1 is complex in the 0.5 to 2.5 ppm area and characteristic of highly saturated compounds like triterpenes (**Table 5**).

The presence of a triplet at 5.24 ppm is characteristic of the trisubstituted ethylenic group (H12) coupled to two protons. The split doublet at 2.85 ppm can be attributed to the proton carried by a tertiary carbon in α of a double bond and coupled to two protons. Finally, in the area from 0.7 to 2.5 ppm seven singletons can be distinguished, proving the presence of 7 methyl groups. This skeleton would be

Compound Vc2: Oleanolic acid		
	Experimental data	**Literature [24]**
Position	δ1H (ppm, CD3OD)	δ1H (ppm, CDCl3)
3	3.15 (1H, dd, J = 11.4 and 4.3. Hz)	3.21 (1H, dd, J = 4.4 Hz and 10.2 Hz)
12	5.24 (1H, t, J = 4.7 Hz)	5.24 (1H, t, J = 3.6 Hz)
18	2.85 (1H, dd, J = 13.6 and 4.3 Hz)	2.82 (1H, dd, J = 4.3 Hz and 12.7 Hz)
23	0.98 (3H, s)	0.96 (3H, s)
24	0.78 (3H, s)	0.78 (3H, s)
25	0.82 (3H, s)	0.84 (3H, s)
26	0.75 (3H, s)	0.76 (3H, s)
27	1.29 (3H, s)	1.25 (3H, s)
29	0.85 (3H, s)	0.87 (3H, s)
30	0.91 (3H, s)	0.93 (3H, s)

Table 5.
Comparison of NMR data1 H of compound Vc2 and oleanolic acid.

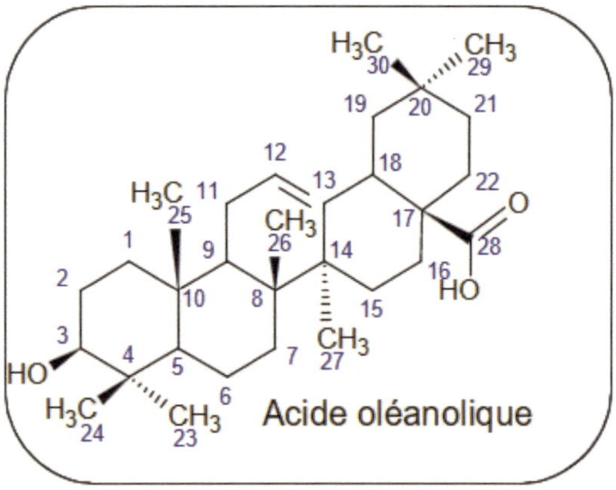

Figure 13.
Molecular structure of oleanolic acid (Acide ursolique) extract from leaves of Vaccinium cereum *var.* cereum.

that of the oleanolic acid (**Figure 13**). However, the spectra being recorded in MeOD, the proton of the acid function could not be distinguished.

The purity of the compound was measured at 70% by integrating the signal from the methyl at position 26 (CH3). The IC_{50} of the mixture with 70% of this compound was determined for α-glucosidase and is 4.2 ± 0.2 μg/mL.

As ursolic acid, oleanolic acid is also present in many plants and found in other species of the genus *Vaccinium* such as *Vaccinium myrtillus*, *Vaccinium macrocarpon*, *V. oxycocos*, *Vaccinium angustifolium* and *Vaccinium vitis-idaea* [25–28].

Oleanolic acid also has several biological activities, such as anti-HIV [37], anticancer prostate [26], liver [30], antioxidant [38], anti-inflammatory [34], hepatoprotective [39] and hypolipidemic [36].

2.3.3 Molecules isolated from the ethyl acetate phase

2.3.3.1 Vc3 compound

The compound Vc3 is a yellowish solid. It has a UV spectrum characteristic of caffeoylquinic acids (λmax ≈ 296 and 326 nm) [40]. The proton and carbon NMR of this compound confirms that it is the methyl ester form of chlorogenic acid with the presence of a methyl at 3.69 ppm and an observed HMBC correlation of this CH3 with C-7 at 174.0 ppm (**Figure 14** and **Table 6**) [41, 42].

2.3.3.2 Vc4 compound

The compound Vc4 is a yellowish solid. It has a UV spectrum characteristic of caffeoylquinic acids (λmax ≈ 296 and 326 nm) [40]. The proton and carbon NMR of

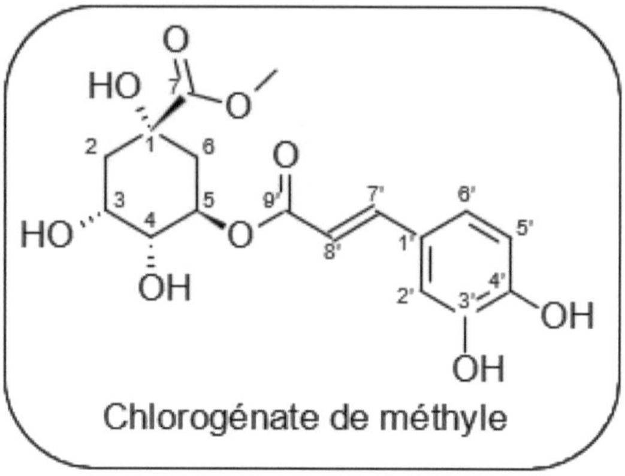

Figure 14.
Molecular structure of methyl chlorogenate (Chlorogénate de méthyle) isolated from leaves of Vaccinium cereum *var.* cereum.

Compound Vc3: Methyl chlorogenate or methyl ester of 5-O (E)-caffeoylquinic acid				
	Experimental data	**Literature [41]**	**Experimental data**	**Literature [41]**
Position	δ1H (ppm, CD3OD)	δ1H (ppm, CD3OD)	δ13C (ppm, CD3OD)	δ13C (ppm, CD3OD)
1			74.5	74.5

Compound Vc3: Methyl chlorogenate or methyl ester of 5-O (E)-caffeoylquinic acid				
	Experimental data	**Literature [41]**	**Experimental data**	**Literature [41]**
2a	2.20 (1H, m)	2.16 (1H, dd, J = 3.8 Hz and 12.3 Hz)	36.6	36.4
2b	2.14 (1H, m)	2.00 (1H, m)	36.6	36.4
3	4.14 (1H, m)	4.17 (1H, m)	68.9	69.0
4	3.74 (1H, m)	3.73 (1H, dd, J = 2.8 Hz and 7.4 Hz)	70.7	70.8
5	5.28 (1H, m)	5.27 (1H, td, J = 4.5 Hz and 7.4 Hz)	70.7	70.4
6a	2.20 (1H, m)	2.21 (1H, dd, J = 2.8 Hz and 12.3 Hz)	36.6	36.7
6b	2.02 (1H, dd, J = 6.2 and 12.5 Hz)	2.02 (1H, m)	36.6	36.7
7			174.0	174.1
1'			126.2	126.3
2'	7.05 (1H, s)	7.03 (1H, d, J = 1.9 Hz)	113.7	114.0
3'			145.4	145.6
4'			148.3	148.4
5'	6.78 (1H, d, J = 8.3 Hz)	6.78 (1H, d, J = 8.2 Hz)	115.2	115.2
6'	6.95 (1H, d, J = 8.3 Hz)	6.95 (1H, dd, J = 1.9 Hz and 8.2 Hz)	121.7	121.6
7'	7.52 (1H, d, J = 15.7 Hz)	7.52 (1H, d, J = 16.1 Hz)	145.8	145.9
8'	6.21 (1H, d, J = 15.7 Hz)	6.21 (1H, d, J = 16.1 Hz)	113.6	113.7
9'			166.9	166.9
O-CH3	3.69 (3H, s)	3.69 (3H, s)	51.7	51.6

Table 6.
Comparison of ^{1}H NMR and ^{13}C NMR data of Vc3 compound and methyl chlorogenate.

this compound is similar to that of compound Vc3, but the absence of the singlet at 3.69 ppm in H NMR1 corresponding to the O-CH3 of the ester indicates the presence of chlorogenic acid (**Figure 15** and **Table 7**) [43].

Both molecules Vc3 and Vc4 did not show α-glucosidase inhibitory activity at 10 μg/mL, but chlorogenic acid and its derivatives are known to have antioxidant [44], bactericidal [45], anti inflammatory, analgesic, antipyretic [46] and anticancer properties by inhibiting carcinogenesis *in vivo* [47].

2.3.3.3 Vc5 compound

The compound Vc5 is a yellowish solid. The 5 aromatic protons H-6, H-8, H-2′, H-3′ and H-6′, present in proton NMR, are characteristic of the quercetin skeleton (**Figure 16**). The presence of a double at 5.25 ppm integrating for a proton (anomeric proton) is characteristic of the presence of an O-glycosidic bond. The chemical shift of

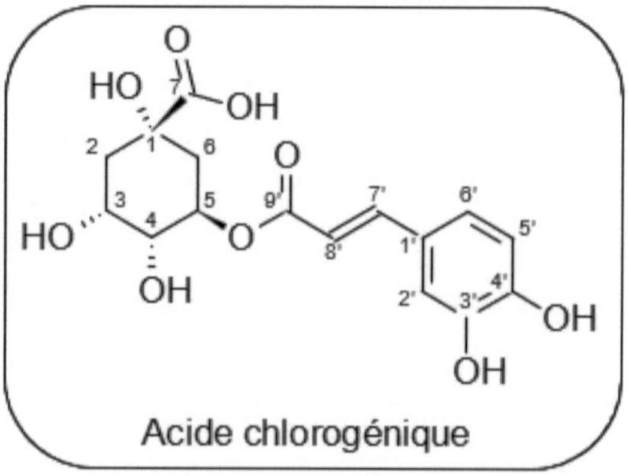

Figure 15.
Molecular structure of chlorogenic acid (Acide chlorogénique) isolated from leaves of Vaccinium cereum *var.* cereum.

Compound Vc4: Chlorogenic acid or 5-O-caffeoylquinic acid				
	Experimental data	**Bibliographic data**	**Experimental data**	**Literature [43]**
Position	δ1H (ppm, CD3OD)	δ1H (ppm, CD3OD)	δ13C (ppm, CD3OD)	δ13C (ppm, CD3OD)
1			75.9	74.7
2a	2.18 (1H, m)	2.17 (1H, dd, J = 3.1 Hz and 14.1 Hz)	38.1	37.3
2b	2.16 (1H, m)	2.04 (1H, dd, J = 4.9 Hz and 14.1 Hz)	38.1	37.3
3	4.14 (1H, m)	4.14 (1H, ddd, J = 3.1 Hz; 3.3 Hz and 4.9 Hz)	70.5	70.5
4	3.74 (1H, dd, J = 2.7 Hz and 7.8 Hz)	3.71 (1H, dd, J = 3.3 Hz and 8.5 Hz)	72.3	72.1
5	5.28 (1H, ddd, J = 4.6 Hz; 7.8 Hz and 9.2 Hz)	5.31 (1H, ddd, J = 4.2 Hz; 8.5 Hz and 9.1 Hz)	72.3	70.6
6a	2.18 (1H, m)	2.14 (1H, dd, J = 4.2 Hz and 14.0 Hz)	38.1	36.8
6b	2.01 (1H, dd, J = 9.2 Hz and 13.7 Hz)	2.08 (1H, dd, J = 9.1 Hz and 14.0 Hz)	38.1	36.8
7			175.1	175.6
1′			127.7	126.4
2′	7.04 (1H, d, J = 2.1 Hz)	7.03 (1H, d, J = 1.9 Hz)	115.2	113.9
3′			146.9	145.3
4′			149.8	148.1
5′	6.78 (1H, d, J = 7.8 Hz)	6.75 (1H, d, J = 8.1 Hz)	116.6	115.1
6′	6.94 (1H, dd, J = 2.1 Hz and 7.8 Hz)	6.94 (1H, dd, J = 1.9 Hz and 8.1 Hz)	123.1	121.6
7′	7.53 (1H, d, J = 15.9 Hz)	7.56 (1H, d, J = 15.8 Hz)	147.3	145.7

Compound Vc4: Chlorogenic acid or 5-O-caffeoylquinic acid				
	Experimental data	**Bibliographic data**	**Experimental data**	**Literature [43]**
8′	6.22 (1H, d, J = 15.9 Hz)	6.22 (1H, d, J = 15.8 Hz)	115.1	113.8
9′			168.4	167.3

Table 7.
Comparison of ^{1}H *NMR and* ^{13}C *NMR data of compound Vc4 and chlorogenic acid.*

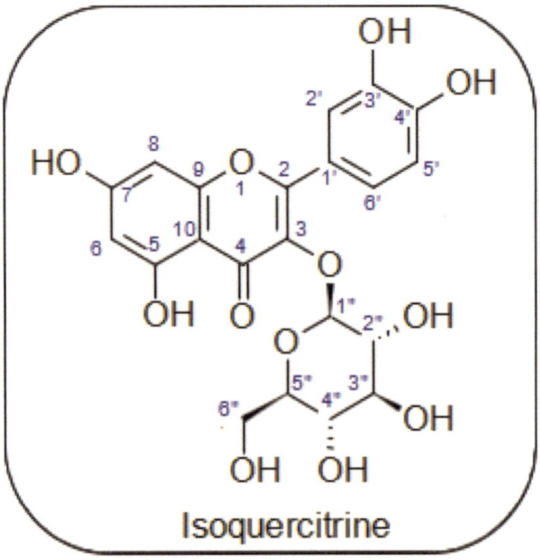

Figure 16.
Molecular structure of isoquercitrin (Isoquercitrine) isolated from leaves of Vaccinium cereum *var.* cereum.

Compound Vc5: quercitrin-3-O-glucoside or isoquercitrin				
	Experimental data	**Bibliographic data**	**Experimental data**	**Literature [48]**
Position	δ1H (ppm, CD3OD)	δ1H (ppm, CD3OD)	δ13C (ppm, CD3OD)	δ13C (ppm, CD3 OD)
2			159.1	159.2
3			135.8	135.8
4			179.6	179.6
5			163.1	163.2
6	6.20 (1H, d, J = 1.7 Hz)	6.20 (1H, s)	100.1	100.1
7			166.3	166.3
8	6.39 (1H, J = 1.7 Hz)	6.39 (1H, s)	94.9	94.9
9			158.7	158.6
10			105.8	105.8
1′			123.2	123.2

Compound Vc5: quercitrin-3-O-glucoside or isoquercitrin				
	Experimental data	Bibliographic data	Experimental data	Literature [48]
2′	7.71 (1H, d, J = 1.7 Hz)	7.70 (1H, s)	117.7	117.7
3′			146.1	146.1
4′			150.0	150.0
5′	6.87 (1H, d, J = 8.5 Hz)	6.87 (1H, d, J = 8.3 Hz)	116.2	116.1
6′	7.59 (1H, dd, J = 1.7 and 8.5 Hz)	7.58 (1H, d, J = 8.3 Hz)	123.2	123.3
1″	5.25 (1H, J = 7.5 Hz)	5.24 (1H, d, J = 7.6 Hz)	104.4	104.5
2″	3.48 (1H, J = 7.9 Hz)	3.48 (1H, t, J = 8.4 Hz)	75.8	75.9
3″	3.43 (1H, J = 8.9 Hz)	3.43 (1H, t, J = 8.9 Hz)	78.2	78.3
4″	3.35 (1H, J = 9.3 Hz)	3.35 (1H, t, J = 9.2 Hz)	71.3	71.4
5″	3.25 (1H, m)	3.22 (1H, m)	78.5	78.5
6 “a	3.58 (1H, dd, J = 5.4 and 11.6 Hz)	3.57 (1H, dd, J = 5.3 Hz and 11.8 Hz)	62.7	62.7
6 “b	3.71 (1H, dd, J = 2.1 and 11.6 Hz)	3.70 (1H, d, J = 11.8 Hz)	62.7	62.7

Table 8.
Comparison of ^{1}H NMR and ^{13}C NMR data of compound Vc5 and isoquercitrin.

the carbons and protons are characteristic of glucose. Compound Vc5 is isoquercitrin (**Table 8**).

Isoquercitrin is a flavonoid found in various medicinal plants. The glycosylated derivative of quercetin would have a bioavailability superior to the aglycone. In addition to the antioxidant effect, it is attributed to various biological activities including antidiabetic, anti-inflammatory during allergic reactions, cardioprotective, and anticancer [49, 50].

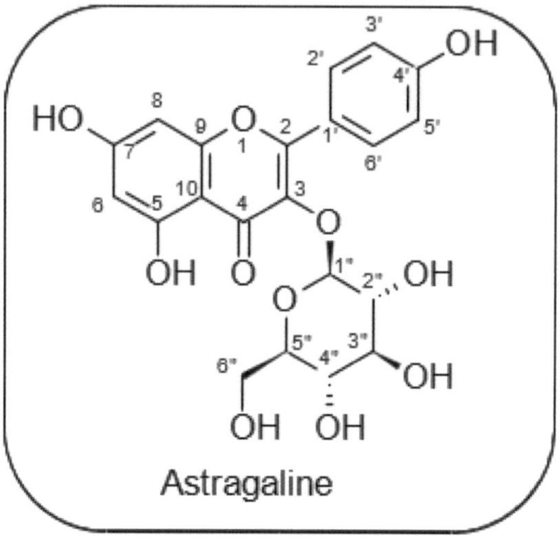

Figure 17.
Molecular structure of astragalin (Astragaline) isolated from leaves of Vaccinium cereum *var.* cereum.

2.3.3.4 Vc6 compound

The compound Vc6 is a yellowish solid. The 6 aromatic protons H-6, H-8, H-2′, H-3' H-5′, and H-6′, present in proton NMR, are characteristic of the kaempferol skeleton (**Figure 17**). The presence of a double at 5.24 ppm integrating for 1 H (anomeric proton) is characteristic of the presence of an O-glycosidic bond. The chemical shift of the carbons and protons is characteristic of glucose. The compound Vc6 is astragalin (**Table 9**).

Compound Vc6: Kaempferol-3-O-glucoside or astragalin				
	Experimental data	**Bibliographic data**	**Experimental data**	**Literature [48]**
Position	δ^1H (ppm, CD_3OD)	δ^1H (ppm, CD_3OD)	δ^{13}C (ppm, CD_3OD)	δ^{13}C (ppm, CD_3 OD)
2			157.1	159.3
3			134	135.8
4			178.1	179.8
5			161.7	163.4
6	6.19 (1H, d, J = 1.9 Hz)	6.17 (1H, d, J = 1.9 Hz)	98.5	100.8
7			164.6	167.9
8	6.39 (1H, d, J = 1.9 Hz)	6.36 (1H, d, J = 1.9 Hz)	93.3	95.5
9			157.7	159.0
10			104.3	105.7
1′			121.36	123.2
2′	8.05 (1H, d, J = 8.6 Hz)	8.05 (1H, d, J = 8.9 Hz)	130.9	132.7
3′	6.88 (1H, d, J = 8.6 Hz)	6.88 (1H, d, J = 8.9 Hz)	114.7	116.5
4′			160.1	162.0
5′	6.88 (1H, d, J = 8.6 Hz)	6.88 (1H, d, J = 8.9 Hz)	114.7	116.5
6′	8.05 (1H, d, J = 8.6 Hz)	8.05 (1H, d, J = 8.9 Hz)	130.9	132.7
1″	5.25 (1H, d, J = 7.3 Hz)	5.22 (1H, d, J = 7.6 Hz)	102.6	104.7
2″	3.44 (1H, dd, J = 7.3 Hz and 9.3 Hz)	3.44 (1H, t, J = 9.2 Hz)	74.3	76.1
3″	3.42 (1H, t, J = 9.1 Hz)	3.41 (1H, t, J = 9.2 Hz)	76.6	78.5
4″	3.33 (1H, m)	3.31 (1H, t, J = 9.8 Hz)	69.9	71.8
5″	3.21 (1H, m)	3.19 (1H, m)	77.0	78.8
6 "*a*	3.53 (1H, dd, J = 5.4 Hz and 11.8 Hz)	3.53 (1H, dd, J = 5.5 Hz and 11.8 Hz)	61.2	63.0
6 "*b*	3.70 (1H, dd, J = 2.2 Hz and 11.8 Hz)	3.68 (1H, dd, J = 2.4 Hz and 11.8 Hz)	61.2	63.0

Table 9.
Comparison of ^{1}H NMR and ^{13}C NMR data of compound Vc6 and astragalin.

Astragalin is a flavonoid endowed with, in addition to its antioxidant effect, an anti-inflammatory activity on atopic dermatitis by acting on the inhibition of inflammation-inducing liposaccharide mediators [48, 51].

3. Conclusion

The study of *Vaccinium cereum* showed a very high interest in the inhibition of α-glucosidase, an enzyme involved in type 2 diabetes (T2DM).

All ethanolic crude extracts from the roots, branches, stems, leaves, flowers, and fruits of *V. c.* var. *cereum*, *V. c.* var. *pubiflorum*, *V. c.* var. *adenandrum* and *V. c.* var. *raiateense,* without exception inhibit this enzyme with low IC_{50} ranging from 0.5 to 9.6 μg/mL compared to acarbose, the positive control used in the treatment of T2DM which inhibits the enzyme with an IC_{50} of 270.5 μg/mL.

The highest activities are observed for roots and branches of three endemic *Vaccinium cereum* taxa with average IC_{50} of 0.8 and 1.7 μg/mL, respectively.

In addition to α-glucosidase inhibitory activity, the extracts showed significant antioxidant and antiradical activities.

We were able to show a positive correlation between the α-glucosidase inhibition test and the antioxidant activities thanks to the accumulated data. The results suggest that the inhibitory activity of α glucosidase is due in particular to the presence of polyphenols.

The phytochemical study of the leaves isolated and identified oleanolic acid and ursolic acid from the dichloromethane phase from the liquid-liquid partition of the ethanolic extract. Both acids showed a low IC_{50} of 10.9 and 4.2 μg/mL on α-glucosidase, respectively, compared to acarbose which has an IC_{50} of 270.5 μg/mL. These two triterpene acids are present in species of the genus *Vaccinium* and have various activities as anticancer, hypolipidemic, hepatoprotective, and antidiabetic.

From the ethyl acetate phase, phenolic compounds were isolated and identified such as chlorogenic acid and its methyl ester, astragalin, and isoquercitrin. In addition to their strong antioxidant activity, these phenolic compounds have anti-inflammatory and antibiotic activities. Hydroxycinnamic acid derivatives are also present in this extract.

Therefore, all these results revealed that this taxon represents a real health benefit and might have promising proprieties to regulate blood sugar.

Acknowledgements

We thank Pr. I. Bombarda (IMBE, Aix-Marseille University) for NMR measurement, the Ministry of Marine Resources, Mines, and Research of French Polynesia for its financial support.

Author details

Sarah Benayad[1], Stéphanie Soulet[2] and Taivini Teai[2]*

1 Higher School of Biological Sciences of Oran (ESSBO), Cité Emir Abdelkader (EX-INESSMO), Oran, Algeria

2 Unité Mixte de Recherche 241 Ecosystèmes Insulaires Océaniens (UMR 241 EIO), Université de Polynésie Française (UPF), Tahiti, French Polynesia

*Address all correspondence to: taivini.teai@upf.pf

References

[1] OMS. Definition, Diagnosis and Classification of Diabetes Mellitus and Its Complications. Geneva: WHO; 1999

[2] Lignes directrices de l'OMS sur l'activité physique et la sédentarité. *Organisation mondiale de la santé*. 2020. ISBN 978-92-4-001486-2

[3] Document de travail de l'OMS: Projet de recommandations visant à renforcer et à suivre les réponses au diabète dans le cadre des programmes nationaux de lutte contre les maladies non transmissibles y compris les cibles potentielles. *Organisation mondiale de la santé*. 2021

[4] Bertrand S, Paupier A, Tuheiava M. Bulletin d'informations sanitaire, épidémiologique et sanitaire. Campagne 2013 de lutte contre le diabète de type 2 en Polynésie française: "Le diabète, je peux l'éviter". Papeete: BISES. 2014; **11**:1-2

[5] Nair SS, Kavrekar V, Mishra A. In vitro studies on alpha amylase and alpha glucosidase inhibitory activities of selected plant extracts. European Journal of Experimental Biology. 2013;**3**:128-132

[6] Pandit R, Phadke A, Jagtap A. Antidiabetic effect of *Ficus religiosa* extract in streptozotocin-induced diabetic rats. Journal of Ethnopharmacology. 2010;**128**:462-466

[7] Lawson-Evi P et al. Activité hypogycémiante de quelques plantes médicinales. Pharmceutical Medical Traditional Africa. 1997;**9**:60-69

[8] Li Y et al. *Punica granatum* flower extract, a potent a-glucosidase inhibitor, improves postprandial hyperglycemia in Zucker diabetic fatty rats. Journal of Ethnopharmacology. 2005;**99**:239-244

[9] Ye F, Shen Z, Xie M. Alpha-glucosidase inhibition from a Chinese medical herb *(Ramulus mori)* in normal and diabetic rats and mice. Phytomedicine. 2002;**9**(2):161-166

[10] Karthic K, Kirthiram KS, Sadasivam S, Thyumanacav B, Palvannan T. Identification of amylase inhibitors from *Syzygium cumini* Linn seeds. IJEB. 2018;**46**:677-680

[11] Delattre J et al. Diabète sucré et stress oxydant. Annales de Biologie Clinique. 1999;**57**:437-444

[12] Helmstadter A, Schuster N. *Vaccinium myrtillus* as an antidiabetic medicinal plant research through the ages. Die Pharmazie. 2010;**65**:315-321

[13] Martineau LC et al. Anti-diabetic properties of the Canadian lowbush blueberry *Vaccinium*. Phytomedecine. 2006;**13**:612-623

[14] Grace H et al. Hypoglycemic activity of a novel anthocyanin-rich formulation from lowbush blueberry, *Vaccinium angustifolium* Aiton. Phytomedecine. 2009;**16**:406-415

[15] Feshani AM, Kouhsari SM, Mohammadi S. *Vaccinium arctostaphylos*, a common herbal medicine in Iran: Molecular biochemical study of its antidiabetic effects on alloxan-diabetic Wistar rats. Journal of Ethnopharmacology. 2011;**133**:67-74

[16] Kianbakth S, Hajiaghaee R. Anti-hyperglycemic effects of *Vaccinium arctostaphylos* L. fruits and leaf extracts in Alloxan-induced diabetic rats. Journal of Medicinal Plants. 2013;**12**:45

[17] Wan C et al. Antioxidant and α-glucosidase inhibitory phenolics

isolated from highbush blueberry flowers. Food Chemistry. 2012;**135**:1929-1937

[18] Matsuo Y et al. Chemical constituents of the leaves of rabbiteye blueberry (*Vaccinium ashei*) and characterisation of polymeric proanthocyanidins containing phenylpropanoid units and A-type linkages. Food Chemistry. 2010;**121**: 1073-1079

[19] Wang SY, Camp MJ, Ehlenfeldt MK. Antioxidant capacity and α-glucosidase inhibitory activity in peel flesh of blueberry. Food Chemistry. 2012;**132**: 1759-1768

[20] Cignarella A et al. Novel lipid-lowering properties of *Vaccinium myrtillus* L. leaves, a traditional antidiabetic treatment, in several models of rat dyslipidaemia: A comparison whith ciprofibrate. Thrombosis Research. 1996;**84**:5

[21] Eid HM et al. Stimulation of AMP-activated protein kinase and enhancement of basal glucose uptake in muscle cells by quercetin and quercetin glycosides, active principles of the antidiabetic medicinal plant *Vaccinium vitis-idaea*. Molecular Nutrition & Food Research. 2010;**54**:991-1003

[22] Grant ML, Fosberg FR, Smith MH. Partial Flora of the Society Islands: Ericaceae to Apocynaceae. Vol. 21. Washington, DC, USA: Smithsonian Institution Press; 1973

[23] Amorati R, Valgimigli L. Advantages and limitations of common testing methods for antioxidants. Free Radical Research. 2014;**49**:633-649

[24] Guvenalp Z et al. Chemical constituents of *Gallium tortumense*. Turkish Journal of Chemistry. 2006;**30**: 515-523

[25] Goetz P. Myrtilles, cassis, anthocyanines et rétino-pathie. Phytothérapie. 2008;**6**:122-124

[26] Kondo M et al. Ursolic acid and its esters: Occurence in cranberries and other Vaccinium fruit and effects on matrix metalloproteinase activity in DU145 prostate tumor cells. Society of Chemical Industry. 2010;**91**:789-796

[27] Booth NL et al. An innovative approach to the safety evaluation of natural products: Cranberry (*Vaccinium macrocarpon* Aiton) leaf aqueous extract as a case study. Food and Chemical Toxicology. 2012;**50**:3150-3165

[28] Szakiel A et al. Triterpenoid content of berries and leaves of bilberry *Vaccinium myrtillus* from Finland and Poland. Journal of Agricultural and Food Chemistry. 2012;**48**:11839-11849

[29] Venugopal R et al. Phytochemicals in diets for breast cancer prevention: The importance of resveratrol and ursolic acid. Food Science and Human Wellness. 2012;**1**:1-13

[30] Yan S-L et al. Oleanolic acid and ursolic acid induce apoptosis in four human liver cancer cell lines. Toxicology. 2010;**24**:842-848

[31] Kim HY, Moon BH, Lee HJ, Choi DH. Flavonol glycosides from the leaves of *Eucommia ulmoides* with glycation inhibitory activity. Journal of Ethnopharmacology. 2014;**93**(2–3): 227-230

[32] Pathak AK et al. Ursolic acid inhibits STAT3 activation pathway leading to suppression of proliferation and chemosensitization of human multiple myeloma cells. Molecular Cancer Research. 2007;**5**:943-955

[33] Balanehru S, Nagarajan B. Protective effect of oleanic acid and ursolic acid

against lipid peroxidation. Biochemistry International. 1991;**24**:981-990

[34] Ovesna Z et al. Pentacyclic triterpenoic acids: New chemoprotective compounds. Neoplasma. 2004;**24**:327-333

[35] Saini P, Gayen P, Kumar D, Nayak A, Mukherjee N, Mukherjee S, et al. Antifilarial effect of ursolic acid from Nyctanthes arbortristis. Molecular and Biochemical Evidences. 2014;**63**: 717-728

[36] Liu J. Pharmacology of oleanolic acid and ursolic acid. Journal of Ethnopharmacology. 1995;**49**:57-68

[37] Zhu Y-M et al. Synthesis and anti-HIV activity of Oleanolic acid derivatives. Bioorganic & Medicinal Chemistry Letters. 2001;**11**:3115-3118

[38] Nickavar B, Gholamreza A. Enzyme assay guided isolation of an α-amylase inhibitor flavonoid from *Vaccinium arctostaphylos* leaves. Iranian Journal of Pharmaceutical Research. 2011;**10**:849-853

[39] Jie L et al. The effect of Chinese hepatoprotective medicines on experimental liver injury in mice. Journal of Ethnopharmacology. 1994;**42**:183-191

[40] Santos-Buelga C, Williamson G. Methods in Polyphenol Analysis. Cambridge, UK: The Royal Society of Chemistry; 2003. p. 383

[41] Silva DB et al. Flavone glycosides from Bidens gardneri Bak. (Asteraceae). Phytochemistry. 2013;**96**:418-422

[42] Chan EWC et al. Coffeolyquinic acids from leaves of *Etlingera* species (zingiberaceae). LWT- Food Science and Technology. 2009;**42**:1026-1030

[43] Yi B et al. Antioxidant phenolic compounds of cassava (*Manihot esculenta*) from Hainan. Molecules. 2010; **16**:10157-10167

[44] Chen JH, Ho C-T. Antioxidant activities of caffeic acid and its related hydroxycinnamic acid compounds. Journal of Agricultural Food Chemistry. 1997;**45**:2374-2378

[45] Lou Z et al. Antibacterial activity and mechanism of action of chlorogenic acid. Food Science. 2011;**76**:398-403

[46] Santos MD dos et al. Evaluation of the anti-inflammatory, analgesic and antipyretic activities of the natural polyphenol chlorogenic acid. Biological and Pharmaceutical Bulletin. 2006;**26**: 2236-2240

[47] Kasaia H et al. Action of chlorogenic acid in vegetables and fruits as an inhibitor of 8-hydroxydeoxyguanosine formation in vitro and in a rat carcinogenesis model. Food and Chemical Toxicology. 2000;**38**:467-471

[48] Kim M-S, Kim S-H. Inhibitory effect of astragalin on expression of lipopolysaccharide-induced inflammatory mediators through NF-κB in macrophages. Archieves in Pharmaceutical Research. 2011;**34**:2101-2107

[49] Valentová K et al. Isoquercitrin: Pharmacology, toxicology, and metabolism. Food and Chemical Toxicology. 2014;**68**:267-282

[50] Rogerio AP et al. Anti-inflammatory activity of quercetin and isoquercitrin in experimental murine allergic asthma. Inflammation Research. 2007;**56**:402-408

[51] Soromou LW et al. Astragalin attenuates lipopolysaccharide-induced inflammatory responses by down-regulating NF-κB signaling pathway. Biochemical and Biophysical Research Communications. 2012;**419**:256-261

Section 4

Benefits of Berry Fruits for Human Health

Chapter 9

Berries: A New Paradigm for Nutraceuticals

Bushra A. Waikar and Pallavi C. Mandave

Abstract

The berries are edible, small, mushy fruit. Different types of berries available throughout the world are strawberry, blueberry, raspberry, mulberry, blackberry, cranberry, gooseberry, elderberry, huckleberry, black current, dewberry, etc. Berries are most commonly rich in phenolic compounds, like flavonoids (i.e. anthocyanins, flavonoids, cathechins), tannins (i.e. ellagitannins, proanthocyanidins), stilbenoids (i.e. resveratrol, piceatannol), phenolic acids (i.e. hydrobenzoic and acid derivatives) and lignans. The growing body of research supports the classification of berries as a functional food with numerous therapeutic and preventative health effects. These organic goods are created to separate the constituents known as flavonoids and anthocyanins. In a number of nutraceutical, pharmacological, medical and cosmetic applications, they are increasingly viewed as an essential component. These compounds showed a wide variety of biological activities through positive effects on the body which includes antioxidant action, control of enzyme activity, and prevention of cellular growth. They all play a role in the regulation of several hormones, including androgens, oestrogens, and thyroids. Consuming diets high in fruits and vegetables is consistently linked to a lower risk of chronic diseases like cancer and cardiovascular disease, according to epidemiological research. In the present review, we aim to assess the health-promoting potential of berries as a pharmaceutical and nutraceutical aspect.

Keywords: berries, chronic diseases, health benefits, nutraceuticals, functional food

1. Introduction

Our planet is affluent in a plant species that hold medicinal qualities. Some of plant varieties have been used since long time for treating various diseases [1]. One of the most significant sources of human diet is plants, particularly the fruits. Plants have been used as a natural source of therapeutic chemicals for countless years. Natural products and foods that promote health have greatly increased in popularity among health experts as a result of ground-breaking developments in the nutritional and medical sciences.

Phytonutrients, phytomedicines and phytotherapy are recent trends that are becoming more and more important in our daily lives [2–4]. They also enhance medicinal advantages and improve immune function to avoid certain diseases with the fewer negative effects [5]. Biological therapies known as nutraceuticals are utilised to promote wellness, stop cancerous processes and fend off symptoms [6]. The understanding that nutraceuticals play a significant role in enhancing health has increased attention on a global scale, and there has been significant reorganisation among the population of the beneficial effects of nutraceuticals in daily life [7]. The perceived health benefits of the nutraceutical goods include lowering the risk of cancer and heart disease as well as preventing hypertension, high cholesterol, being overweight and osteoporosis. It also showed positive/therapeutical effect in some disease such as diabetes, arthritis, cataracts, macular degeneration (which causes permanent blindness), menopausal symptoms, sleeplessness and poor memory.

Nutraceuticals are now being recognised as one of the methods for stopping these disorders [8–13]. Fruits are the most accessible and affordable source of nutrients, including vital amino acids, carbs, proteins, vitamins and minerals [14, 15]. In addition to all of these, they are useful for treating a variety of illnesses. Fruits and vegetables provide our body with a wide range of nutrients and phytochemicals that are essential for optimum health thanks to their various chemical compositions and colorations. Consuming fruits regularly is associated with a reduced chance of developing cancer, and the fruits' phytochemicals' synergistic and additive actions are what give them their powerful antioxidant and anticancer properties [1]. They have a significant amount of bioactive ingredients [7]. Berries are among the fruits that contain a variety of antioxidant phenolics, including flavonoids, stilbenes, tannins and phenolic acids. Free radicals and reactive oxidant species are produced in a wide variety of physiological processes [16].

In fruits, berries are the main source of phenolic compounds, which also include tannins, phenolic acids and flavonoids (flavonols, anthocyanins, isoflavonoids, flavanols and flavones). Berries may be used to treat a variety of diseases pharmacologically by acting on inflammation and oxidative stress, which are primarily the main causes of diabetes, cancer, neurological, cardiovascular and other ailments [17].

Thus, the presence of pharmacologically active compounds for the treatment of diabetes, cardiovascular issues and various other diseases, and this chapter examines commonly consumed berries like strawberries, mulberries, blackberries, blueberries, cranberries and raspberries as well as their polyphenols as potential medicinal foods.

2. Berry cultivation: international and national scenario

Berries are known for their unique aroma, sweet taste and high phenolic contents. This helps to improve the standard of the diet. Strawberries, blueberries, cranberries and black and red raspberries are the most popular fruit kinds. **Figure 1** shows an illustration of a few of the berries. Generally speaking, they range in coloration from red to purple or black depending on the species. Berries are used in a variety of food products, such as jams, jellies and drinks, in addition to being eaten fresh or dried.

Since environmental elements during cultivation are crucial for affecting fruit components and consequently the final quality, many of them have their own distinctive zones that are ideal for growing. High concentrations of phenolic components, such as flavonoids, phenolic acids, tannins, stilbenes and lignans, are present in all variations. Many people's eating habits have altered as a result of growing knowledge about the beneficial health effects of food bioactives and efforts to maximise nutritional value.

Figure 1.
Different berries cultivated around the world.

2.1 International scenario

Various berries are cultivated in various parts of worlds. **Table 1** shows various berries along with scientific names and their origin in the world.

2.2 Berries in India

Most of the strawberry produced in India comes from Mahabaleshwar, Maharashtra, where 85 per cent of the crop is farmed. In and around the Mahabaleshwar region, large-scale production of gooseberry, raspberry and mulberry is also practised [19]. The Maharashtra area of Nashik is where grapes are primarily farmed. The greatest grape-producing city in India is Nashik, which is also referred to as the 'Wine Capital of India'. It is at the top of the list of Maharashtra regions that grow grapes, followed by Sangli, Satara and Ahmednagar. Maharashtra produces 62.7 per cent of the total grapes in India. Mulberry is mainly produced in Karnataka, Andhra Pradesh and Tamil Nadu states, which accounts for a total of 90 per cent production in India. It is also produced in West Bengal, Himachal Pradesh and the north-eastern states.

Gooseberry's largest producer in India is Uttar Pradesh state, with a 35 per cent share in total India production. Tamil Nadu (28%) is the second largest gooseberry-producing state followed by Madhya Pradesh (14%). Mysore is a strong, upright plant that produces an abundance of delicious, purple-black raspberries. **Table 2** shows various berry-producing regions in India.

Sr. No	Common name	Scientific name	Origin
1.	Strawberry	*Fragaria ananassa*	Brittany, France
2.	Blueberry	*Vacciniumcyano coccus*	North America
3.	Raspberry	*Rubusidaeus*	Turkey
4.	Blackberry	*Rubussubg. Rubus*	Armenia
5.	Cranberry	*Vaccinium subgenus Oxycoccus*	North-eastern North America
6.	Boysenberry	*Rubusursinus x Rubusidaeus*	Anaheim, California
7.	Lingonberry	*Vacciniumvitis-idaea*	Native to the Scandinavian region of Northern Europe
8.	Elderberry	*Sambucus*	Europe
9.	Huckleberry/ Bilberry	*Vaccinium membranaceum Dougl*	The Andes and other mountainous regions of South America, Eastern North America
10.	Goji Berry/ Wolfberry	*Lyciumbarbarum*	China
11.	Black Mulberry	*Morusnigra*	Western Asia (Iran and Afghanistan)
12.	Black Currant	*Ribesnigrum*	Northern Europe
13.	Gooseberry	*Ribesuva-crispa*	North-eastern and north-central United States and the adjacent regions of Canada
14.	Açai Berry	*Euterpeoleracea*	Northern Brazil
15.	Hardy Kiwi/Kiwi Berry/Siberian Gooseberry	*Actinidia arguta*	Russian Far East Native to China, Korea and Japan
16.	Salmonberry	*Rubusspectabilis*	Western U.S.A. and Canada
17.	Saskatoon Berry/ Juneberry	*Amelanchieralnifolia*	Western Canada
18.	Cloudberry	*Rubuschamaemorus*	North of Sweden
19.	Bearberry	*Arctostaphylosuva-ursi*	South-western China
20.	Red Mulberry	*Morusrubra*	Eastern and Central North America
21.	Caperberry	*Capparisspinosa*	Native of Mediterranean from Canary Islands and Morocco to Crimea and Armenia
22.	Chokeberry	*Aronia*	Eastern Canada and the Central and the Eastern United States
23.	Chokecherry	*Prunusvirginiana*	North America
24.	Red Currant	*Ribesrubrum*	Western Europe
25.	Dewberry	*Rubusflagellaris*	North America and Northern Europe

The information is adapted from Pire et al. [18].

Table 1.
Berries and their origin country.

Sr. No	Common name	Scientific name	Cultivated in India
1.	Strawberry	*Fragaria ananassa*	Bengal, Delhi, Haryana, Himachal Pradesh, Uttar Pradesh, Maharashtra, West Punjab and Rajasthan
2.	Mulberry	*Morusalba*	Andhra Pradesh, Himachal Pradesh Karnataka, Tamil Nadu, West Bengal and the North-eastern states
3.	Grapes	*Vitisvinifera*	Maharashtra, Karnataka, Tamil Nadu and Mizoram
4.	Gooseberry	*Phyllanthus emblica*	West Bengal, north India and eastern India
5.	Raspberry	*Rubusidaeus*	Mysore
6.	Kanta Berry (Kanntam)	*Carissa carandas*	Konkan region
7.	Zaraberry	*Phyllantha ceaeacidus*	In and around Goa
8.	Cranberry	*Vaccinium subgenus Oxycoccus*	Andhra Pradesh, Tamil Nadu, Kerala, Maharashtra, Karnataka and West Bengal

The information is adapted from https://nhb.gov.in/report_files/strawberry/STRAWBERRY.htm [20].

Table 2.
Berries cultivation in India.

3. Berries: a brimful of secondary metabolites

Till date, many reviews and studies have been reported on berries consumption and health [21]. Presently, many berries are used as components in dietary supplements and different functional foods [22]. Berries are much credible in various phytochemicals and nutrients, which are reported to enhance health and avert various chronic diseases.

Phenolic compounds consist of flavonoids, that is flavonols, flavones, flavanols, flavanones, isoflavonoids, anthocyanins and tannins; phenolic acids are the primary phytochemicals present in berries. Many reports and studies calliper the health effects of anthocyanins that are a class of flavonoids [23]. Anthocyanidin, an aglycon component of anthocyanins including pelargonidin, cyanidin, peonidin, delphini 17 petunidin and malvidin, and a sugar component make up these pigments, which are widely present in berries. An excellent source of anthocyanins is strawberries. The glycosidic derivatives of pelargonidin and cyanidin are the anthocyanins that are most frequently detected in strawberries [24].

In addition, berries contain stilbenes, hydrolyzable tannins known as gallic and ellagic acid esters, and condensed (nonhydrolyzable) tannins known as proanthocyanidins. It contains a range of vitamins, dietary fibres and minerals that are all necessary. It is low in calories and fats and plentiful in sugars [17]. Vitamin C, dietary fibres, potassium and folates are all present in berries like raspberries, blackberries and black currants. These berries range in vitamin C content from 9.7 to 60 mg/100 g [25]. Strawberries, blackberries and raspberries are excellent providers of potassium and folate (vitamin B9). Blackberries contain the lowest levels of vitamin C, whereas

Sr.No	Berry	Secondary metabolites	References
1.	Strawberry	Flavonoids **a. Flavonols** Quercetin-glucuronide, Quercetin-3-malonyglucoside, Quercetin-3-glucuronide Quercetin-rutinoside, Quercetin-glucoside, Fisetin Kaempferol-3-malonyglucoside, Kaempferol-glucuronide, Kaempferol-3 glucoside Kaempferol-coumaroyl-glucoside **b. Flavanols** Proanthocyanidin B1; Proanthocyanidin B3 (þ)-Catechin; Proanthocyanidin trimer **c. Anthocyanins** 3-malonylglucoside; 3-malonylglucosyl-5-glucoside); Cyanidin and their derivatives (−3-glucoside; 3-rutinoside; −3-glucoside; −3-acetylglucoside; Pelargonidin and its derivatives (−3-galactoside; −3-rutinoside; −3-arabinoside; −3-malylglucoside; −3-malonylglucoside; −3,5-diglucoside; −disaccharide (hexose þ pentose) acylated with acetic acid); 5-Pyranopelargonidin-3-glucoside	[29, 32, 33]
2.	Mulberry	Volatile glycoside; Moracin – M; 2-Arylbenzofuran; Inorganic anions; Cyanidin 3-rutinoside; Flavonoids; Chlorogenic acid and Rutin; 1-Deoxynojirmycin; Anthocyanins; Hydrophobic flavonoids; Moracin – C; Cynadin-3-Glucoside; Melatonin (N-acetyl-5-methoxytryptamine	[34–45]
3.	Grapes	Anthocyanins and proanthocyanidins Epicatechin; Epicatechin; Catechin; Gallic acid; Gallate Protocatechuic acid; Procyanidin B1; Procyanidin B2; Procyanidin B4; Protocatechic acid	[46, 47]
4.	Gooseberry	**Polyphenols** Alkaloids; flavonoids; caffeic acid; rutin; phenolic acids; glucocorticoids; gallic acid; ellagic acid; withanolides; mangiferin; vitamins; saponins; tannins; carotenoids	[48–66]
5.	Raspberries	**Anthocyanins-** Cyanidin and its derivatives (−3-sophoroside; 5-diglucoside; 2Gglucosylrutinoside; −3-glucoside; −3-rutinoside); Pelargonidin and its derivatives (−3-sophoroside; −3-(2G-glucosylrutinoside); −3-glucoside; −3-rutinoside) **Ellagitannins** Hydroxycinnamic acids—caffeic, and ferulic acids Hydroxybenzoic acids—ellagic acid *p*-hydroxybenzoic acids; *p*-coumaric **Flavonols** (free and conjugated form] quercetin and kaempferol **Condensed tannins**	[67–74]
6.	Cranberry	**Anthocyanins**—Arabinosides of peonidin and cyanidin and Galactosides; anthocyanins, Flavonol; glycosides, proanthocyanidins; Organic acids; Phenolic acids	[75, 76]
7.	Chokeberry	Cyanidin 3-glycosides; Chlorogenic acid	[77]
8.	Red current	**Phenolic compounds**—Protocatechuic; chlorogenic acid; ellagic acid; rutin; gallic acid; catechin; syringic acid; vanillic acid; p-coumaric acid; o-coumaric acid; phloridzin and ferulic acid; caffeic acid; quercetin. **Organic acids**—Citric acid; tartaric acid; malic acid; succinic acid; and fumaric acid.	[78]

Table 3.
Berries along with their secondary metabolite composition.

strawberries have the highest levels. Blackberries and blueberries contain high levels of vitamin K, while cranberries are rich in vitamin E [26, 27]. Beta-carotene, lutein and zeaxanthin are all rich in blackberries [28]. Among these berries, black currants have the highest concentration of calcium, iron, phosphorus and potassium [28].

Nearly 40 phenolic compounds, including kaempferol, quercetin, pelargonidin, cyanidin and ellagic acid glycosides, as well as flavanols, p-coumaric 36 derivatives and ellagitannins, are found in strawberries. This complete identification of the phenolic compounds found in these fruits is provided by this study [29].

Fresh mulberry fruit contains carbs in the form of simple sugars, starch, soluble and insoluble fibres, as well as starch. Mulberry fruit has a high water content and a low calorie count. The two nutrients most commonly present in these fruits are iron and vitamin C. These fruits include adequate levels of potassium, vitamin E and vitamin K. Additionally, it contains a good deal of anthocyanins, which give fruits their colour and have advantageous health effects. According to recent investigations, mulberry leaves contain a significant number of bioactive substances, primarily alkaloids, flavonoids, y-aminobutyric acid (GABA) and phenolic acids. 1-Deoxynojirimycin (DNJ) is one of the bioactive substances that can be discovered in mulberries [30].

Anthocyanin flavonoids make up to 60% of the total polyphenolics in a ripe blueberry. Thus, anthocyanins contribute mainly to the health advantages of blueberries [31]. Both flavonoids and non-flavonoids are included in blueberries' polyphenolic components. Proanthocyanidins and flavonols are two classes of flavonoids that can be found in blueberries. The hydroxycinnamic acid esters, particularly chlorogenic acid, are common non-flavonoid polyphenolic chemicals found in blueberries [31]. The different secondary metabolites found in the berries are shown in **Table 3**. The role of strawberry metabolites is in a range of disorders.

4. Strawberry metabolites in various diseases

4.1 Diabetes

Diabetes sharply raises the risk of atherosclerosis, a cardiovascular condition that causes diabetes individuals to die more frequently. The development of vascular disease is significantly influenced by high glucose-induced vascular inflammation and subsequent endothelial dysfunction in diabetes [79, 80]. Pro-inflammatory cytokines, hyperglycemia, dyslipidemia, vascular adhesion molecules and diabetes all contribute to vascular dysfunction by raising endothelial cell (EC) inflammation. Treatment of these diabetes-related complications frequently places a significant financial burden on both the individuals and society at large. Finding affordable medications to treat these problems is necessary due to the current diabetes epidemic [81, 82].

There is evidence that several of the phenolic quercetin components found in strawberries, such as ellagic acid and catechin in their pure forms, have antidiabetic properties [83]. Through the modification of hepatic enzyme activity, glucose metabolism and a lipid profile, flavonoids contribute to the aetiology of diabetes and its consequences [84]. In the streptozotocin-induced diabetic rat model, rutin stimulates the production and translocation of GLUT4, which promotes glucose transfer to soleus muscle tissue [83–85]. Fruit flavonoids control diabetes with no adverse consequences by maintaining cellular homeostasis [86].

4.2 Myocardial infarction

Heart attack is another common name for myocardial infarction. It is a very serious disorder that develops when the heart muscle is not getting enough blood flow. However, blockages in one or more of the heart's arteries are most frequently to blame for the lack of blood flow.

According to recent research, eating two to three servings of strawberries each week reduces the incidence of myocardial infarction in people [87]. In rats, flavone and apigenin exhibit a protective effect against MIRI. It significantly lowers the level of malondialdehyde (MDA) and enhances the activity of superoxide dismutase (SOD) in MIRI. This suggests that apigenin inhibits free radical oxidation and increases the activity of oxidase in tissue 7, which may provide cardiac protection [88]. Epicatechin pretreatment can lessen the increase in metalloproteinase in the myocardial infarction area, demonstrating the ability of flavonoids to inhibit metalloproteinase activity in MIRI [89].

4.3 Postprandial inflammation

The rise in blood glucose that happens after eating is known as postprandial. Consuming strawberries has been found to lessen postprandial inflammation in obese persons who have a high-carb, moderate-fat meal. In adults with cardiovascular risk factors, strawberry consumption lowers vascular adhesion molecules. Digestive enzymes and intestinal flora completely metabolise anthocyanins in humans, which raise the possibility that the circulating metabolite may be responsible for the vascular effects [90].

4.4 Antioxidant activity

Antioxidant activity is the restriction or prevention of food oxidation, including that of proteins, lipids and carbohydrates [91]. Out of all the fruits that people eat, strawberries contribute the most to cellular antioxidant activity [92].

Wogonin, a 7 O-methylated flavone, has been shown to have cardiac protective effects by reducing irreversible I/R injury and significant ischemia-induced arrhythmia, which are related to antioxidant capacity and anti-inflammatory activity [93].

4.5 Antihyperglycemic and antihypertensive

High blood glucose is a frequent name for hyperglycemia. As a result of the body not producing enough insulin, there is too much sugar in the blood. Brazilian strawberries have antihyperglycemic and antihypertensive benefits, according to the World Health Organisation, which defines hypertension as the condition in which the blood pressure is excessively high. They demonstrate the ellagic acid derivatives' inhibitory action against a-amylase and a-glucosidase as well as the enzyme convertant angiotensin I, which may be used to treat hypertension and hyperglycemia [94].

4.6 Obesity and glycaemia

A disorder known as obesity involves an unregulated accumulation of bodily fat. It is a medical issue that raises the risk of other illnesses and health issues, including heart disease, diabetes, high blood pressure and occasionally certain cancers

(Mayo Clinic staff). It is not just a cosmetic issue. It has been demonstrated that a powder prepared from freeze-dried strawberries can lower obesity and enhance glycaemic control. Strawberries include anthocyanins, which can lower blood sugar levels [95].

4.7 Oxidative stress and inflammation

The fluctuation between the introduction and accumulation of oxygen reactive species (ROS) in cells and tissues results in an occurrence known as oxidative stress [94]. By enhancing translocation and GLUT4 activity, decreasing oxidative stress and inflammation, and increasing translocation, strawberries' polyphenol content can improve glucose metabolism and peripheral glucose absorption in insulin-sensitive tissue [96]. Inflammation is root cause of the many diseases. Antioxidant plays an important role in the neutralization of the reactive oxygen species (ROS). The berries are rich source of the antioxidants. Hence, the berries have potential nutraceuticals/bioactive compounds. These compounds should be explored to open the way for medicinal foods against various diseases.

5. Conclusions

One of the most popular meal items is fruit. Berry fruits specifically play a significant role in the diet due to its high nutritional content, which includes vitamins, minerals, phytochemicals and phenolic compounds. The health-promoting effects of berry fruits on a number of ailments are obviously enforced in this chapter, and these effects are linked to their phenolic components. On the other hand, there might be variations in these phenolic contents based on the type and variety of fruit as well as other elements including the environment and agricultural conditions. Many human bodily systems, including the gastrointestinal, cardiovascular, immunological and nervous systems, benefit greatly from the majority of berries. In summary, it is clear that berries are very healthy for people due to their high levels of flavonoids, anthocyanins and total phenolic acids, which have strong antioxidant properties. Reactive oxygen species (ROS), which are the main cause of many diseases, can be inhibited by polyphenols, which have a remarkable level of antioxidant capability. Berry polyphenols therefore have a credible potential to reduce the negative health impacts of ROS-induced illnesses and disorders.

Acknowledgements

The authors are grateful to the Yashavantrao Chavan Institute of Science, Karmveer Bhaurao Patil University, Satara, India, for providing us the opportunity to work.

Conflict of interest

The authors declare no conflict of interest.

Author details

Bushra A. Waikar and Pallavi C. Mandave*
Department of Biotechnology, Yashavantrao Chavan Institute of Science, Karmveer Bhaurao Patil University, Satara, India

*Address all correspondence to: mandavepallavi@gmail.com

References

[1] Mahima AK, Rahal A, Deb R, Latheef SK, Samad HA. Immunomodulatory and therauptic potential of herbal, traditional/ indigenous and ethanoveterinary medicine. Pakistan Journal of Biological Sciences. 2012;**15**:754-774

[2] Bland JS. Phytonutrition, phytotherapy and phytopharmacology. Alternative therapies. Health And Medicine. 1996;**2**:73-76

[3] Berger MM, Shenkin A. Vitamins and trace elements: Practical aspects of supplementation. Nutrition. 2006;**22**:952-955

[4] Bagchi D. Nutraceuticals and functional foods regulations in the United States and around the world. Toxicology. 2006;**221**:1-3

[5] Ramaa CS, Shirode AR, Mundada AS, Kadam VJ. Nutraceuticals-an emerging era in the treatment and prevention of cardiovascular diseases. Current Pharmaceutical Biotechnology. 2006;**7**:15-23

[6] Brower V. Nutraceuticals: Poised for a healthy slice of the healthcare market? Nature Biotechnology. 1998;**16**:728-731

[7] Dolkar D, Bakshi P, Wali VK, Sharma V, Shah RA. Fruits as nutraceuticals. Ecology, Environment and Conservation. 2017;**23**(2):113-118

[8] Heyland DK. In search of the magic nutraceuticals: Problems with current approaches. Journal of Nutrition. 2001;**131**(9):2591-2595

[9] Klein C, Sato T, Meguid MM, Miyata G. From food to nutritional support to specific nutraceuticals: A journey across time in the treatment of disease. Journal of Gastroenterology. 2000;**35**:1-6

[10] Whitman M. Understanding the perceived need for complementary and alternative nutraceuticals: Lifestyle issues. Clinical Journal of Oncology Nursing. 2001;**5**:190-194

[11] Elizabeth AC. Over the counter products: Nonprescription medications, nutraceuticals, and herbal agents. Clinical Obstetrics and Gynecology. 2002;**45**(1):89-98

[12] Kalra EK. Nutraceutical definition and introduction. AAPS PharmSciTech. 2003;**5**(3):25

[13] Sengupta A, Ghosh S, Das S. Tomato and garlic can modulate azoxymethane-induced colon carcinogenesis in rats. European Journal of Cancer Prevention. 2003;**12**:195-200

[14] Murphy MM, Barraj LM, Herman D, Bi X, Cheatham R, Randolph RK. Phyto-nutrient intake by adults in the United States in relation to fruit and vegetable consumption. Journal of the Academy of Nutrition and Dietetics. 2012;**11**:222-229

[15] Bumgarner N, Scheerens R, Kleinhenz MD. Nutritional yield: A proposed index for fresh food improvement illustrated with leafy vegetable data. Plant Foods for Human Nutrition. 2012;**67**:215-222

[16] Paredes-López O, Cervantes-Ceja ML, Vigna-Pérez M, Hernández-Pérez T. Berries: Improving human health and healthy aging, and promoting quality life—A review. Plant Foods for Human Nutrition. 2010;**65**(3):299-308

[17] Golovinskaia O, Wang CK. Review of functional and pharmacological activities of berries. Molecules. 2021;**26**(13):3904

[18] Pire T. 25 Different Types of Berries and Why you Should be Eating each and every One of them. PureWow; 2022

[19] Bhosale V, Mullaji N. Profitability and swot analysis of strawberry production. 2021. Available from: https://nhb.gov.in/report_files/strawberry/STRAWBERRY.htm

[20] Sawant S, Lee B, Janghoon S, Seo H. The Indian gooseberry (Emblica officinalis) industry and cultivation in India. Journal of the Korean Society of International Agriculture. 2022;**34**:199-204

[21] Manic C, Scalbert A, Morand C, Rémésy C, Jiménez L. Polyphenols: Food sources and bioavailability. The American Journal of Clinical Nutrition. 2004;**79**:727-747

[22] Tungmunnithum D, Thongboonyou A, Pholboon A, Yangsabai A. Flavonoids and other phenolic compounds from medicinal plants for pharmaceutical and medical aspects: An overview, 2018. Medicine. 2018;**5**(3):93

[23] Zhang G, Celli B, Brooks MS. Chapter 1: Natural sources of anthocyanins. In: Anthocyanins from Natural Sources: Exploiting Targeted Delivery for Improved Health. Cambridge: Royal Society of Chemistry; 2019. pp. 1-33

[24] Smeriglio A, Barreca D, Bellocco E, Trombetta D. Proanthocyanidins and hydrolysable tannins: Occurrence, dietary intake and pharmacological effects. British Journal of Pharmacology. 2017;**174**(11):1244-1262

[25] Elen Khachatrian, 2021. Nutrition & Microbiology at YSU.

[26] Dismore MI, Haytowitz DB, Gebhardt SE, Peterson JW, Booth SL. Vitamin K content of nuts and fruits in the US diet. Research and Professional Briefs. **103**(12):1650-1652

[27] Golovinskaia O, Wang C-K. Review of functional and pharmacological activities of berries. Molecules. 2021;**26**(13):3904

[28] Aaby K, Ekeberg D, Skrede G. Characterization of phenolic compounds in strawberry (Fragaria× ananassa) fruits by different HPLC detectors and contribution of individual compounds to total antioxidant capacity. Journal of Agricultural and Food Chemistry. 2007;**55**(11):4395-4406

[29] Jan B, Parveen R, Zahiruddin S, Khan MU, Mohapatra S, Ahmad S. Nutritional constituents of mulberry and their potential applications in food and pharmaceuticals: A review. Saudi Journal of Biological Sciences. 2021;**28**(7):3909-3921

[30] Kalt W, Cassidy A. Luke R Howard, Robert Krikorian, April J stull, Francois Tremblay, Raul Zamora-Ros; recent research on the health benefits of blueberries and their Anthocyanins. Advances in Nutrition. 2020;**11**(2):224-236

[31] Arai Y, Watanabe S, Kimira M, Shimoi K, Mochizuki R, Kinae N. Dietary intakes of flavonols, flavones and isoflavones by Japanese women and the inverse correlation between quercetin intake and plasma LDL cholesterol concentration. The Journal of Nutrition. 2000;**130**:2243-2250

[32] Basu A, Rhone M, Lyons TJ. Berries: Emerging impact on cardiovascular health. Nutrition Reviews. 2010;**68**(3):168-177

[33] Andallu B, Varadacharyulu N. Control of hyperglycemia and retardation of cataract by mulberry (Morus indica L.) leaves in streptozotocin diabetic rats Ind.

The Journal of Experimental Biology. 2002;**40**:791-795

[34] Singab AN, El-Beshbishy HA, Yonekawa M, Nomura T, Fukai T. Hypoglycemic effect of Egyptian Morus alba root bark extract: Effect on diabetes and lipid peroxidation of streptozotocin-induced diabetic rats. Journal of Ethnopharmacology. 2005;**100**(3):333-338

[35] Chen MG, Jin PH, Huang LX. Energy analysis of mulberry-silkworm ecosystem in China. Chinese Journal of Applied Ecology. 2006;**17**:233-236

[36] Kang TH, Hur JY, Kim HB, Ryu JH, Kim SY. Neuroprotective effects of the cyanidin-3-O-beta-d-glucopyranoside isolated from mulberry fruit against cerebral ischemia. Neuroscience Letters. 2006;**391**(3):122-126

[37] Fallon E, Zhong L, Furne JK, Levitt MD. A mixture of extracts of black and green teas and mulberry leaf did not reduce weight gain in rats fed a high-fat diet. Alternative Medicine Review. 2008;**13**(1):43-49

[38] Hunyadi AI, Herke K, Veres A, Erdei A, Simon GT. Volatile glycosides from the leaves of Morus alba with a potential contribution to the complex anti-diabetic activity. Natural Product Communications. 2014;**9**:145-147

[39] Bazylak GM, A. Siepak and Gryn; content of some inorganic anions in aqueous brews of white mulberry leaves used as components of dietary supplements for diabetic subjects. Planta Medica. 2014;**80**:1476

[40] Natic MM, Dabic DC, Papetti A, Fotiric MM, Ognjanov AV, Ljubojevic M, et al. Analysis and characterization of phytochemicals in mulberry (*Morus alba* L.) fruits grown in Vojvodina, North Serbia. Food Chemistry. 2015;**171**:128-136

[41] Choi JW, Synytsya A, Capek P, Bleha R, Pohl R, Park YI. Structural analysis and anti-obesity effect of a pectic polysaccharide isolated from Korean mulberry fruit Oddi (*Morus alba* L.). Carbohydrate Polymers. 2016;**146**:187-196

[42] Wang C, Yin LY, Shi XY, Xiao H, Kang K, Liu XY, et al. Effect of cultivar, temperature, and environmental conditions on the dynamic change of melatonin in mulberry fruit development and wine fermentation. Journal of Food Science. 2016;**81**:M958-M967

[43] Xiao HX, Lingli Z, Xueling G. Phenolic compounds content and antioxidant activity of mulberry wine during fermentation and aging. American Journal of Food Technology. 2017;**12**:367-373

[44] Kiran T, Yuan YZ, Andrei M, Fang Z, Jian GZ, Zhao JW. 1-Deoxynojirimycin, its potential for management of non-communicable metabolic disorders, trends food. Science and Technology. 2019;**89**:88-99

[45] He F, Liang N, Mu L, Pan Q, Wang J, Reeves MJ, et al. Anthocyanins and their variation in red wines I. monomeric anthocyanins and their color expression. Molecules. 2012;**17**:1571-1601

[46] Rodríguez Montealegre R, Romero Peces R, Chacón Vozmediano JL, Martínez Gascueña J, García Romero E. Phenolic compounds in skins and seeds of ten grape *Vitis vinifera* varieties grown in a warm climate. Journal of Food Composition and Analysis. 2006;**19**:687-693

[47] Qiu L, Zhao F, Zhi-Hu Jiang Z-H, Chen L-X, Zhao Q, Liu H-X, et al.

Steroids and flavonoids from Physalis alkekengi var. franchetii and their inhibitory effects on nitric oxide production. Journal of Natural Products. 2008;**71**(4):642-646

[48] Lim TK. Physalis angulate. Physalis peruviana. In: Edible Medicinal and Non-Medicinal Plants: Volume 6, Fruits. Dordrecht: Springer Science and Business Media; 2013. pp. 283-299 pp. 300-309

[49] Rengifo-Salgado E, Vargas-Arana G. Physalis angulata L. (Bolsa Mullaca): A review of its traditional uses, chemistry and pharmacology. Boletín Latinoamericano y del Caribe de Plantas Medicinales y Aromáticas. 2013;**12**(5):431-445

[50] Men RZ, Li N, Ding WJ, Hu ZJ, Ma ZJ, Cheng L. Unprecedent aminophysalin from Physalis angulata. Steroids. 2014;**88**:60-65

[51] Li X, Zhao J, Yang M, Liu Y, Li Z, Li R, et al. Physalins and withanolides from the fruits of *Physalis alkekengi* L. var. franchetii (mast.) Makino and the inhibitory activities against human tumor cells. Phytochemistry Letters. 2014;**10**:95-100. DOI: 10.1016/j.phytol.2014.08.004

[52] Kusumaningtyas RW, Laily N, Limandha P. Potential of ciplukan (Physalis angulata L.) as source of functional ingredient. Procedia Chemistry. 2015;**14**:367-372

[53] Meira CS, Guimarães ET, Santos JAF, Moreira DRM, Nogueira RC, Tomassini TCB, et al. *In vitro* and *in vivo* antiparasitic activity of Physalis angulata L. concentrated ethanolic extract against Trypanosoma cruzi. Phytomedicine. 2015;**22**:969-974

[54] Liu X-G, Jiang F-Y, Gao P-Y, Jin M, Yang D, Nian Z-F, et al. Optimization of extraction conditions for flavonoids of Physalis alkekengi var. franchetii stems by response surface methodology and inhibition of acetylcholinesterase activity. Journal of the Mexican Chemical Society. 2015;**59**(1):59-66

[55] Pinto LA, Meira CS, Villarreal CF, Vannier-Santos MA, de Souza CVC, Ribeiro IM, et al. A seco-steroid from Physalis angulata L., has immunosuppressive activity in peripheral blood mononuclear cells from patients with HTLV1-associated myelopathy. Biomedicine & Pharmacotherapy. 2016;**79**:129-134

[56] Bahmani M, Rafieian-Kopaei M, Naghdi N, Nejad ASM, Afsordeh O. Physalis alkekengi: A review of its therapeutic effects. Journal of Chemical and Pharmaceutical Sciences. 2016;**9**(3):1472-1475

[57] Fukushima A, Nakamura M, Suzuki H, Yamazaki M, Knoch E, Mori T, et al. Comparative characterization of the leaf tissue of Physalis alkekengi and Physalis peruviana using RNA-seq and metabolite profiling. Frontiers in Plant Science. 2016;**7**:1883 12 pages

[58] Cobaleda-Velásco M, Alanís-Bañuelos RE, Almaraz AN, Rojas-López M, González-Valdez LS, Ávila-Reyes JA, et al. Phenolic profiles and antioxidant properties of *Physalis angulata L.* as quality indicators. Journal of Pharmacy and Pharmacognosy Research. 2017;**5**(2):114-128

[59] Rivera DE, Ocampo YC, Castro JP, Barrios L, Diaz F, Franco LA. A screening of plants used in Colombian traditional medicine revealed the antiinflammatory potential of Physalis angulata calyces. Saudi Journal of Biological Sciences. 2018;**26**(7):1758-1766

[60] Svobodova B, Kuban V. Solanaceae: A family wellknown and still surprising.

In: Petropoulos SA, Ferreira ICFR, Barros L, editors. Phytochemicals in Vegetables: A Valuable Source of Bioactive Compounds. Bentham Science Publishers; 2018. pp. 296-372

[61] Tuan AHL, Thao DT, Dung DT, Kiem PV, Quang TH, Hai Yen PT, et al. Phytochemical constituents and cytotoxic activity of Physalis angulate L. growing in Vietnam. Phytochemistry Letters. 2018;**27**:193-196

[62] Li AL, Chen BJ, Li GH, Zhou MX, Li YR, Ren DM, et al. Physalis alkekengi L. var. franchetii (mast.) Makino: An ethnomedical, phytochemical and pharmacological review. Journal of Ethnopharmacology. 2018;**210**:260-274

[63] Lv H, Fu C-S, Hu H-X, Wang X-N, Ren D-M, Lou H-X, et al. Chemical constituents from the calyxes of Physalis alkekengi L. var. franchetii (mast.) Makino. Biochemical Systematics and Ecology. 2018;**78**:63-65

[64] Wen X, Hempel J, Schweiggert RM, Ni Y, Carle R. Carotenoids and carotenoid esters of red and yellow physalis (Physalis alkekengi L. and P. pubescens L.) fruits and calyces. Journal of Agricultural and Food Chemistry. 2017;**65**(30):6140-6151

[65] Wen X, Erşan S, Li M, Wang K, Steingass CB, Schweiggert RM, et al. Physicochemical characteristics and phytochemical profiles of yellow and red physalis (*Physalis alkekengi* L. and *P. pubescens* L.) fruits cultivated in China. Food Research International. 2019;**120**:389-398. DOI: 10.1016/j.foodres.2019.03.002

[66] Mullen W, Lean ME, Crozier A. Rapid characterization of anthocyanins in red raspberry fruit by high-performance liquid chromatography coupled to single quadrupole mass spectrometry. Journal of Chromatography. A. 2002;**966**:63-70

[67] Gu L, Kelm MA, Hammerstone JF, Beecher G, Holden J, Haytowitz D, et al. Concentrations of proanthocyanidins in common foods and estimations of normal consumption. The Journal of Nutrition. 2004;**134**:613-617

[68] Beekwilder J, Jonker H, Meesters P, Hall RD, van der Meer IM, Ric de Vos CH. Antioxidants in raspberry: On-line analysis links antioxidant activity to a diversity of individual metabolites. Journal of Agricultural and Food Chemistry. 2005;**53**:3313-3320

[69] Kassim A, Poette J, Paterson A, Zait D, McCallum S, Woodhead M, et al. Environmental and seasonal influences on red raspberry anthocyanin antioxidant contents and identification of quantitative traits loci (QTL). Molecular Nutrition & Food Research. 2009;**53**:625-634

[70] Määttä-Riihinen KR, Kamal-Eldin A, Torronen AR. Identification and quantification of phenolic compounds in berries of Fragaria and Rubus species (family Rosaceae). Journal of Agricultural and Food Chemistry. 2004;**52**:6178-6187

[71] Remberg SF, Sonsteby A, Aaby K, Heide OM. Influence of postflowering temperature on fruit size and chemical composition of Glen ample raspberry (Rubus idaeus L.). Journal of Agricultural and Food Chemistry. 2010;**58**:9120-9128

[72] Jakobek L, Seruga M, Novak I, Medvidovic-Kosanovic M. Flavonols, phenolic acids and antioxidant activity of some red fruits. Dtsch Lebensm Rundsch. 2007;**103**:369-378

[73] Clifford MN, Scalbert A. Ellagitannins—Nature, occurrence and

dietary burden. Journal of the Science of Food and Agriculture. 2000;**80**:1118-1125

[74] Fuleki T, Francis FJ. Quantitative methods for anthocyanins. Purification of cranberry anthocyanins. Journal of Food Science. 1968;**33**:266-269

[75] Seeram NP, Adams LS, Hardy ML, Heber D. Total cranberry extract versus its phytochemical constituents: Antiproliferative and synergistic effects against human tumor cell lines. Journal of Agricultural and Food Chemistry. 2004;**52**(9):2512-2517

[76] Takahashi A, Shimizu H, Okazaki Y, Sakaguchi H, Taira T, Suzuki T, et al. Anthocyanin-rich phytochemicals from Aronia fruits inhibit visceral fat accumulation and hyperglycemia in high-fat diet-induced dietary obese rats. Journal of Oleo Science. 2019;**64**(12):1243-1250

[77] Ersoy N, Kupe M, Gundogdu M, Ilhan G, Ercisli S. Phytochemical and antioxidant diversity in fruits of currant (Ribes spp.). Notulae Botanicae Horti Agrobotanici Cluj-Napoca. 2018;**46**:381-387

[78] Tabit CE, Chung WB, Hamburg NM, Vita JA. Endothelial dysfunction in diabetes mellitus: Molecular mechanisms and clinical implications. Reviews in Endocrine & Metabolic Disorders. Mar 2010;**11**(1):61-74. DOI: 10.1007/s11154-010-9134-4. PMID: 20186491; PMCID: PMC2882637

[79] Petrie JR, Guzik TJ, Touyz RM. Diabetes, hypertension, and cardiovascular disease: Clinical insights and vascular mechanisms. The Canadian Journal of Cardiology. 2018;**34**(5):575-584

[80] Chistiakov DA, Orekhov AN, Bobryshev YV. Endothelial barrier and its abnormalities in cardiovascular disease. Frontiers in Physiology. 2015;**6**:365

[81] Toma L, Stancu CS, Sima AV. Endothelial dysfunction in diabetes is aggravated by glycated lipoproteins; novel molecular therapies. Biomedicines. 2020;**9**(1):18

[82] Al-Ishaq RK, Abotaleb M, Kubatka P, Kajo K, Büsselberg D. Flavonoids and their anti-diabetic effects: Cellular mechanisms and effects to improve blood sugar levels. Biomolecules. 2019;**9**(9):430. DOI: 10.3390/biom9090430

[83] Kreft S, Knapp M, Kreft I. Extraction of rutin from buckwheat (Fagopyrum esculentum Moench) seeds and determination by capillary electrophoresis. Journal of Agricultural and Food Chemistry. 1999;**47**:4649-4652. DOI: 10.1021/jf990186p

[84] Ghorbani A. Mechanisms of antidiabetic effects of flavonoid rutin. Biomedicine & Pharmacotherapy. 2017;**96**:305-312. DOI: 10.1016/j.biopha.2017.10.001

[85] Tanveer A, Akram K, Farooq U, Hayat Z, Shafi A. Management of diabetic complications through fruit flavonoids as a natural remedy. Critical Reviews in Food Science and Nutrition. 2017;**57**(7):1411-1422

[86] Cassidy A, Mukamal KJ, Liu L, Franz M, Eliassen AH, Rimm EB. High anthocyanin intake is associated with a reduced risk of myocardial infarction in young and middle-aged women. Circulation. 2013;**127**:188-196

[87] Cheng XT, Bai JP, Zhang HZ, Yu KM. The protective effects of apigenin in myocardium of rats with ischemia reperfusion injury. Pharmacology and Clinics of Chinese Materia Medica. 2011;**2**:48-51

[88] Yamazaki KG, Romero-Perez D, Barraza-Hidalgo M, Cruz M, Rivas M, Coetez-Gomez B, et al. Short and long term effects of (−)-epicatechin on myocardial ischemia reperfusion injury. American Journal of Physiology Heart & Circulatory Physiology. 2008;**295**:761-767

[89] Edirisinghe I, Banaszewski K, Cappozzo J, Sandhya K, Ellis CL, Tadapaneni R, et al. Strawberry anthocyanin and its association with postprandial inflammation and insulin. The British Journal of Nutrition. 2011;**106**:913-922

[90] Preedy VR, Watson RR, editors. Olives and Olive Oil in Health and Disease Prevention. Second ed. Academic Press; 2021

[91] Kelly L, Wolfe XK, He X, Dong M, Zhang Q, Liu RH. Cellular antioxidant activity of common fruits. Journal of Agricultural and Food Chemistry. 2008;**56**(18):8418-8426

[92] Lee YM, Cheng PY, Chen SY, Chung MT, Sheu JR. Wogonin suppresses arrhythmias, inflammatory responses, and apoptosis induced by myocardial ischemia/reperfusion in rats. Journal of Cardiovascular Pharmacology. 2011;**58**(2):133-142

[93] Pinto MDS, Kwon YI, Apostolidis E, Lajolo FM, Genovese MI, Shetty K. Functionality of bioactive compounds in Brazilian strawberry (*Fragariax ananassa* Duch.) cultivars: Evaluation of hyperglycemia and hypertension potential using in vitro models. Journal of Agricultural and Food Chemistry. 2008;**56**(12):4386-4392

[94] Djurica D, Holt RR, Ren J, Shindel AW, Hackman RM, Keen CL. Effects of a dietary strawberry powder on parameters of vascular health in adolescent males. The British Journal of Nutrition. 2016;**116**:639-647

[95] Pizzino G, Irrera N, Cucinotta M, Pallio G, Mannino F, Arcoraci V, et al. Oxidative stress: Harms and benefits for human health. Oxidative Medicine and Cellular Longevity. 2017;**2017**:8416763

[96] Paquette M, Medina AS, Weisnagel LSJ, Desjardins Y, Marois J, Pilon G, et al. Strawberry and cranberry polyphenols improve insulin sensitivity in insulin-resistant, non-diabetic adults: A parallel, double-blind, controlled and randomised clinical trial. British Journal of Nutrition. 2017;**117**(4):519-531

Chapter 10

Cranberry Effects in Urinary Tract Infections

Gabriela Cimadon, Sabrina Barili, Alana da Silva and Beatriz Bonadiman

Abstract

Cranberry is scientifically known as *Vaccinium macrocarpon*, belongs to the family of Ericaceae, and grows in humid environments, such as forests and swamps in North American countries. It is widely used by Native American peoples and by Europeans as a food and also in traditional medicine for health benefits such as cardioprotective and anticarcinogenic properties and prevention of stomach ulcers. Some studies have demonstrated the potential against bacteria that reach the urinary tract and cause infections. According to the research by Salo, the fruit is composed of 88% water, acid organic, fruits, vitamin C, flavonoids, catechins, and anthocyanidins. It can be found in the form of dried fruits, extracts, and juices, which usually contain only 10 to 25% cranberry concentrate. In addition, it was revealed that anthocyanidins and the proanthocyanidins present in cranberry are tannins with a natural defense function against microorganisms, such as those responsible for urinary tract infections (UTI). This berry has a potential curative and preventive effect for various diseases, especially urinary tract infection. Therefore, it becomes relevant to clarify the effect of this natural product on urinary tract infections.

Keywords: natural product, treatment, infections, cranberry, urinary tract infections

1. Introduction

The American cranberry is one of the main native fruit cultures from North America. The so-called marvelous fruit, which belongs to the *Ericaceae* family and the species *Vaccinium macrocarpon*, has recently become the center of attention due to its significant therapeutic potential [1] and being a unique source that is rich in various bioactive classes, including flavonoids, anthocyanins, type A proanthocyanidins, polyphenols, carotenoids, and vitamins C and E [2].

This plant's extract has been cultivated mainly for use in the industry as a meat preservative. In the medical industry, its main use is for wound healing, as well as to treat dental cavities and stomach ulcers, to prevent bacterial infections of the urinary tract caused by *E. coli*, and is likely to have anticancer properties. Several varieties of fungal pathogens are also susceptible to cultivated cranberry [3].

The daily consumption of a variety of fruit is recommended to keep healthy nutrition standards, to meet a suitable intake of micronutrients, and to promote the intake of a diversity of phytochemicals. Wild berries, including cranberries, represent a rich source of phenolic bio-actives that can contribute to human health [2].

Although they are not usually eaten raw, cranberries are available in a wide range of choices including juice and sauce, as well as dried fruit in cereal bars, cheese, chocolate, and other confectionary. All these choices of product are beneficial to health in a similar way and are easy to include in daily meals. Moreover, cranberry powder and extracts are now used in food and dietary supplements [2].

2. Effects of cranberry

Cranberry products have been used for centuries in popular medicine for the urinary tract health, and their antioxidant and antibacterial properties have been long reported [4, 5]. Cranberry's biological activities are attributed to a diversified phytochemical profile with flavonoids such as flavonols, anthocyanins, and proanthocyanidins and catechins, phenolic acids, and triterpenoids [6].

The cranberry fruit has a vast class of phenolic compounds, responsible for its antioxidant activities. Chlorogenic acid and quercetin found in cranberry fruit are associated with a reduction in blood pressure, obesity, dyslipidaemia, and the effects of vasodilation improvement dependent on the endothelium [7–9].

Triterpenoids found in the wax of the cranberry fruit peel are associated with anti-inflammatory, anti-tumor, and anticancer activities [10, 11]. Previous studies have reported the efficacy of cranberry extracts to inhibit the growth and induction of apoptosis against different lineages of cancer cells, such as lung, ovarian, breast, prostate, and cervical cancer [4, 12, 13]. These antitumorigenic properties can be related with its antioxidant activity, which attenuates oxidative stress through the modulation of endogen enzymes and the reduction of oxidative markers [3, 14].

The polyphenols, anthocyanins, and proanthocyanidins found in cranberries act in the cell signaling pathways responsible for inflammation [4, 15]. The anti-inflammatory activity of cranberry products has been shown in the modulation of inflammatory markers of different inflammatory diseases such as colitis [16], hepatic inflammation [15], inflammatory intestinal disease [17], and pancreatic inflammation [14]. The administration of cranberry capsules standardized with proanthocyanidins has shown effect on the occurrence and severity of cystitis by radiation in prostate cancer patients [4].

Cranberries have been associated with the urinary tract health. Previous studies suggest that cranberry reduces urinary tract infections through negative regulation and interference on multiple virulence factors, which are associated with stress tolerance and microorganism survival in the urinary tract [18]. Cranberry products showed action against motility of the pathogens *Pseudomonas aeruginosa*, *Escherichia coli*, and *Proteus mirabilis*. Furthermore, a modulation of *quorum sensing* has been shown, which is a signaling pathway that influences several other bacterial virulence factors [19–21].

Bioactive compounds with selective antiviral activity in pathogenic bacteria can be extremely useful in the fight against bacterial infections caused by pathogens resistant to broad-spectrum antibiotics [20]. It has been shown that cranberries and propolis act to protect against adherence, motility, biofilm, and bacterial multiplication in the urinary tract when used as an alternative in the treatment for urinary tract infections by *E. coli* [22].

A number of studies have shown the potential of this species for pharmacological application directed to oral health. In this context, cranberry hydroalcoholic extracts inhibited the growth of six oral pathogenic microorganisms on the cariogenic, periodontal, endodontic, and candida order, namely, *Streptococcus mutans*, *Lactobacillus acidophilus*, *P. gingivalis*, *Aggregatibacter actinomycetemcomitans*, *E. faecalis*, and *C. albicans* [23, 24].

Cranberry products are associated with benefits to digestive health. They contain fibers and polyphenols related to the beneficial modulation in the gut microbiota, promoting the development of beneficial microorganisms, decreasing the occurrence of pathogenic microorganisms, and modulating the production of microbial metabolic products that help digestive health, such as short-chain fatty acids (SCFA) [16, 25, 26].

Proanthocyanidins are associated to the benefits against *Helicobacter pylori* infections, and the daily intake of cranberries is indicated for at least eight weeks [27].

Clinical studies with humans using cranberry products have focused mainly on the health of the urinary and cardiovascular tracts, oral health, and gastrointestinal epithelium [14] and have shown the high biotechnological potential of this species; however, it is still underestimated regarding other biological activities and therapeutic and pharmacological applications.

3. Cranberry effects on the treatment of urinary tract infections

Urinary tract infections (UTIs) are the most disseminated bacterial infections, and they represent a heavy economic and medical burden worldwide [28]. UTI is a pathology that predominates in all populations, even though women are the most affected. Approximately 50 to 70% of women have at least one episode of UTI during their lives, and 20 to 30% have recurring episodes [29].

In the majority of the infections of cystitis, 85%, the main etiological agent is *E. coli* (UPEC), but other microorganisms cause infections; Gram-negative bacteria *Klebsiella pneumoniae* and gram-positive coccus such as staphylococcus and enterococcus seem to be implicated in these infections [30, 31].

In Latin America, high percentages of *E. coli* strains were found to be resistant to fluoroquinolones (17.5–18.9%), ampicillin (61.4%), amoxicillin, clavulanic acid (18.6%), ceftriaxone (20.5%), gentamycin (25.0%), and trimethoprim/sulfamethoxazole (45%) [32].

It is important to point out that uropathogens can form biofilm to more easily adhere and increase their resistance. This affects the remaining microorganisms, causing higher adherence and multiplication, which may result in thick biofilms that significantly reduce the efficacy of antimicrobial treatments. When biofilm resources are limited, mature bacteria detach and can colonize a new surface to repeat the cycle [33].

Conventionally, the use of antibiotics to treat this pathology has been effective but may lead to resistance among uropathogens and other adverse side effects such as damage to gut microbiota. For these reasons, there is increasing interest in the search for natural therapies to prevent and treat UTIs to fight against the increase of bacterial resistance to antibiotics and the high recurrence rates [31, 34].

In these circumstances, one of the most studied alternative therapies against UTIs is the treatment with *V. macrocarpon* [32]. Salo et al. [35] have brought to light that anthocyanidins and proanthocyanidins found in cranberries are tannins with a natural defense function against microbial cells, and are a source of high-quality antioxidants.

Several clinical studies suggest that the intake of cranberry juice or cranberry supplements can reduce the occurrence of UTIs in healthy women [36, 37]. These observations were supported by the results of *in vitro* research, which show that compounds derived from cranberry (polyphenols and type A proanthocyanidins, for example) can interfere with bacterial adherence to epithelial cells of the urinary tract, attenuate the uropathogenic reservoir in the gastrointestinal tract, and suppress the inflammatory cascade [38, 39].

Among other possible mechanisms behind the protective effects of cranberries against UTIs is the ability of cranberry polyphenols to act as antiadhesive agents to prevent/inhibit the adherence of pathogens to uroepithelial cell receptors, which seems to be an important step in the pathogenesis of these infections [40].

As an alternative for the treatment of UTIs caused by *E. coli*, Ranfaig et al. [22] presented a study that assessed the levels of genome expression of this bacteria and measured them in the presence of cranberry and propolis products separately, and later compared with the non-treated isolate. They observed that cranberry along with propolis acted favoring protection against adherence, motility, biofilm, bacterial multiplication, and virulence of the urinary tract. Furthermore, the authors affirmed that the administration of both compounds was able to make the propolis intensify the action of cranberry, thus reaching the desired therapeutic effect.

Saramarasinghe et al. [41] performed a study with an isolate of *E. coli* CTXM-15, chosen due to its fast dissemination and resistance to some groups of antibiotics. It was submitted for analysis by qRT-PCR (quantitative reverse transcription-polymerase chain reaction) of the behavior of virulence factors after treatment with Cysticlean®, a cranberry product containing around 240 mg of proanthocyanidins.

The ten selected genes performed significant roles in normal cell functions such as iron absorption, toxin production, and stress survival in *E. coli* resistant to third-generation cephalosporin, and its relative genic expression was determined after exposure to Cysticlean® using qRT-PCR. Results showed that Cysticlean® significantly reduces the expression of a large variety of genes responsible for aptitude and virulence [41].

Two complementary mechanisms were suggested to explain the ability of American cranberry to reduce urinary tract infections. The first mechanism was credited to its biocide activity, which is made possible by the production of various elements that can damage the uropathogenic bacteria. Other mechanism to its antiviral activity through the negative regulation and interference with others virulence factors associated to stress tolerance and survival in the urinary tract [18].

Ventura [32] has shown the report of a patient who has suffered from recurring urinary infections for 20 years and made use of oral Ciprofloxacin for three days, later finding out they had *E. coli* following urine culture and resistance to the antibiotic they were using. Therefore, it was suggested to the patient to have treatment with cranberry extract.

The treatment described by Ventura [32] consisted of the intake of 5 g of the fruit, the equivalent amount to 18 mg of proanthocyanidins, dissolved in 250 mL of water, twice a day for 20 days, with follow-up attention. Six days after the end of treatment, the patient was asymptomatic, urine culture was normal, and antibiogram was negative. Thirteen weeks later, the patient was still asymptomatic, and urine cultures remained negative regarding bacterial growth.

4. Conclusion

The information obtained through the studied papers allows to conclude that the use of *V. macrocarpon* is an effective therapy in cases of urinary tract infections and has prophylactic action in cases of recurrent infections due to proanthocyanidins and anthocyanidins, which impair bacterial adherence to the urinary tract wall, and antiviral action that causes bacterial weakening and inactivation.

Finally, this natural product is promising for the treatment of urinary infections caused by gram-negative and -positive bacteria.

Author details

Gabriela Cimadon[1], Sabrina Barili[2], Alana da Silva[3] and Beatriz Bonadiman[2]*

1 University of West Santa Catarina, Xanxere, SC, Brazil

2 Community University of Chapecó, Chapecó, SC, Brazil

3 Federal University of Santa Catarina, Florianópolis, SC, Brazil

*Address all correspondence to: beadasilvarosa@gmail.com

References

[1] Jurikova T, Skrovankova S, Mlcek J, Balla S, Snopek L. Bioactive Compounds, Antioxidant Activity, and Biological Effects of European Cranberry (*Vaccinium oxycoccos*) . 2018;**24**:24. https://www.ncbi.nlm.nih.gov/pmc/articles/PMC6337168/

[2] Blumberg JB, Camesano TA, Cassidy A, Etherton PK, Howel A, Manach C, et al. Cranberries and their bioactive constituents in human health. Advances in Nutrition. 2013;**4**:618-632. DOI: 10.3945/an.113.004473

[3] Rasool M, Malik A, Abdul Basit Ashraf M, Mubbin R, Ayyaz U, Waquar S, et al. Phytochemical analysis and protective effects of Vaccinium macrocarpon (cranberry) in rats (Rattus norvegicus) following ethylene oxide-induced oxidative insult. Bioengineered. 2021;**12**:4593-4604. DOI: 10.1080/21655979.2021.1955528

[4] Hamilton K, Bennett NC, Purdie G, Herst PM. Standardized cranberry capsules for radiation cystitis in prostate cancer patients in New Zealand: A randomized double blinded, placebo controlled pilot study. Supportive Care in Cancer. 2015;**23**:95-102. DOI: 10.1007/s00520-014-2335-8

[5] Das S. Natural therapeutics for urinary tract infections—A review. Future Journal of Pharmaceutical Sciences. 2020;**2020**:6. DOI: 10.1186/s43094-020-00086-2

[6] Jurikova T, Skrovankova S, Mlcek J, Balla S, Snopek L. Bioactive compounds, antioxidant activity, and biological effects of European cranberry (vaccinium oxycoccos). Molecules. 2019;**2019**:24. DOI: 10.3390/molecules24010024

[7] Sedbare R, Silina D, Janulis V. Evaluation of the Phytochemical Composition of Phenolic and Triterpene Compounds in Fruit of Large Cranberries (*Vaccinium macrocarpon* Aiton) Grown in Latvia Plants (Basileia). 2022;**11**(20): 2725. https://www.ncbi.nlm.nih.gov/pmc/articles/PMC9609109/

[8] Galvez J, Zevallos L, Velazquez D. Chlorogenic acid: Recent advances on its dual role as a food additive and a nutraceutical against metabolic syndrome. Molecules. 2017;**22**:358 https://www.ncbi.nlm.nih.gov/pmc/articles/PMC6155416/

[9] Pourmasoumi M, Hadi A, Najafgholizadeh A, Joukar F, Ghanaei F. The effects of cranberry on cardiovascular metabolic risk factors: A systematic review and meta-analysis. Clinical Nutrition. 2020;**39**:774-788. https://pubmed.ncbi.nlm.nih.gov/31023488/

[10] Oszmiański J, Kolniak-Ostek J, Lachowicz S, Gorzelany J, Matlok N. Phytochemical compounds and antioxidant activity in different cultivars of cranberry (*Vaccinium Macrocarpon* L). Journal of Food Science. 2017;**82**:2569-2575. https://pubmed.ncbi.nlm.nih.gov/28973819/

[11] Wu X, Xue L, Tata A, Song M, Neto C, Xiao H. Bioactive components of polyphenol-rich and non-polyphenol-rich cranberry fruit extracts and their Chemopreventive effects on colitis-associated Colon Cancer. Journal of Agricultural and Food Chemistry. 2020. https://pubmed.ncbi.nlm.nih.gov/32390426/

[12] Wang Y, de B. Harrington P, Chen P. Analysis of phenolic compositions in

cranberry dietary supplements using UHPLC-HRMS. Journal of Food Composition and Analysis. 2020;**2020**:86. DOI: 10.1016/j.jfca.2019.103362

[13] Weh KM, Zhang Y, Howard CL, Howell AB, Clarke JL, Kresty LA. Cranberry polyphenols in Esophageal Cancer inhibition: New insights. Nutrients. 2022;**2022**:14. DOI: 10.3390/nu14050969

[14] Santana DG, Oliveira AS, Souza MTDS, Santos JTDC, Hassimotto NMA, Silva AMDOE, et al. Vaccinium macrocarpon Aiton extract ameliorates inflammation and hyperalgesia through oxidative stress inhibition in experimental acute pancreatitis. Evidence-based Complementary and Alternative Medicine. 2018. DOI: 10.1155/2018/9646937

[15] Glisan SL, Ryan C, Neilson AP, Lambert JD. Cranberry extract attenuates hepatic inflammation in high-fat-fed obese mice. Journal of Nutritional Biochemistry. 2016;**37**:60-66. DOI: 10.1016/j.jnutbio.2016.07.009

[16] Cai X, Han Y, Gu M, Song M, Wu X, Li Z, et al. Dietary cranberry suppressed colonic inflammation and alleviated gut microbiota dysbiosis in dextran sodium sulfate-treated mice. Food & Function. 2019;**10**:6331-6341. DOI: 10.1039/c9fo01537j

[17] Xiao X, Kim J, Sun Q, Kim D, Park CS, Lu TS, et al. Preventive effects of cranberry products on experimental colitis induced by dextran sulphate sodium in mice. Food Chemistry. 2015;**167**:438-446. DOI: 10.1016/j.foodchem.2014.07.006

[18] Othman M. Is cranberry effective in preventing recurrent urinary tract infection in women. IOSR Journal of Pharmacy. 2016;**2016**:7-15

[19] Hidalgo G, Chan M, Tufenkji N. Inhibition of Escherichia coli CFT073 flic expression and motility by cranberry materials. Applied and Environmental Microbiology. 2011;**77**:6852-6857. DOI: 10.1128/AEM.05561-11

[20] Maisuria VB, Los Santos YL, Tufenkji N, Déziel E. Cranberry-derived proanthocyanidins impair virulence and inhibit quorum sensing of Pseudomonas aeruginosa. Scientific Reports. 2016;**2016**:6. DOI: 10.1038/srep30169

[21] McCall J, Hidalgo G, Asadishad B, Tufenkji N. Cranberry impairs selected behaviors essential for virulence in Proteus mirabilis HI4320. Canadian Journal of Microbiology. 2013;**59**:430-436. DOI: 10.1139/cjm-2012-0744

[22] Ranfaing J, Dunyach-Remy C, Louis L, Lavigne JP, Sotto A. Propolis potentiates the effect of cranberry (Vaccinium macrocarpon) against the virulence of uropathogenic Escherichia coli. Scientific Reports. 2018;**2018**:8. DOI: 10.1038/s41598-018-29082-6

[23] Philip N, Walsh LJ. Cranberry polyphenols: Natural weapons against dental caries. Dental Journal (Basel). 2019;**2019**:7. DOI: 10.3390/dj7010020

[24] Kumar V, Ankola A, Sankeshwari R, Jalihal S, Deepak V, Jois HS. Assessment of the antimicrobial efficacy of hydroalcoholic fruit extract of cranberry against Socransky complexes and predominant cariogenic, mycotic and endodontic climax communities of the Oral cavity: An extensive in-vitro study. Journal of Oral and maxillofacial. Pathology. 2019;**23**:407-411. DOI: 10.4103/jomfp.JOMFP_3_18

[25] Bekiares N, Krueger CG, Meudt JJ, Shanmuganayagam D, Reed JD. Effect of

sweetened dried cranberry consumption on urinary proteome and fecal microbiome in healthy human subjects. OMICS. 2018;**22**:145-153. DOI: 10.1089/omi.2016.0167

[26] Khoo C, Duysburgh C, Marzorati M, van den Abbeele P, Zhang D. A freeze-dried cranberry powder consistently enhances SCFA production and lowers abundance of opportunistic pathogens in vitro. Biotech. 2022;**2022**:11. DOI: 10.3390/biotech11020014

[27] Li Z, Ma J, Guo Y, Liu W, Li M, Zhang L, et al. Suppression of *helicobacter pylori* infection by daily cranberry intake: A double-blind, randomized, placebo-controlled trial. Journal of Gastroenterology and Hepatology. 2021;**36**:927-935. https://www.ncbi.nlm.nih.gov/pmc/articles/PMC8246812/

[28] Montorsi F, Gandaglia G, Salônia A, Briganti A, Mirone V. Effectiveness of a combination of cranberries, lactobacillus rhamnosus, and vitamin C for the Management of Recurrent Urinary Tract Infections in women: Results of a pilot study. European Urology. 2016;**70**:912-915. https://www.sciencedirect.com/science/article/abs/pii/S0302283816302500

[29] Simoes L, Souza L. Avaliação in vitro da atividade antibacteriana de sucos de cranberry (Vaccinium macrocarpon) sobre cepas de *Escherichia coli* responsáveis por infecção urinária. Maringá PR. 2018. https://docs.bvsalud.org/biblioref/2019/11/1024973/rbacvol-51-2-2019-ref-745.pdf

[30] Flores-Mireles AL, Walker JN, Caparon M, Hultgren SJ. Urinary tract infections: epidemiology, mechanisms of infection and treatment options. Nature Reviews. Microbiology. 2015;**13**:269-284. https://www.ncbi.nlm.nih.gov/pmc/articles/PMC4457377/

[31] Tamadonfar KO, Omattage NS, Spaulding CN, Hultgren SJ. Reaching the end of the line: Urinary tract infections. Microbiol. espectro. 2019;**7**:1-16. https://journals.asm.org/doi/10.1128/microbiolspec.BAI-0014-2019?url_ver=Z39.88-2003&rfr_id=ori:rid:crossref.org&rfr_dat=cr_pub%20%200pubmed

[32] Ventura J. Infecção do trato urinário por *Escherichia coli* resistente a antibióticos tratados com *Vaccinum macrocarpon*: relato de caso. http://rpmi.pe/ojs/index.php/RPMI/article/view/19/1

[33] Jagannathan V, Viswanathan P. Proanthocyanidins—Proanthocyanidins—Will they effectively restrain conspicuous bacterial strains devolving on urinary tract infection? ? Journal of Basic Microbiology. 2018;**58**:567-578. https://onlinelibrary.wiley.com/doi/10.1002/jobm.201800131

[34] Beerepoot M, Geerlings S. Non-antibiotic prophylaxis for urinary tract infections . 2016; 5:36. https://www.ncbi.nlm.nih.gov/pmc/articles/PMC4931387/

[35] Salo J, Uhari M, Helminen M, Korpi M, Nieminem J, Pokka T, Kontiokari T. Cranberry Juice for the Prevention of Recurrences of Urinary Tract Infections in Children: A Randomized Placebo-Controlled Trial. 2012. https://academic.oup.com/cid/article/54/3/340/302662?login=false

[36] Maki KC, Kaspar KL, Khoo C, Derrig LE, Filho AL, Gupta K. Consumption of a cranberry juice beverage lowered the number of clinical urinary tract infection episodes in women with a recent history of urinary tract infection. American Journal of Clinical Nutrition. 2016;**103**:1434-1442. https://academic.oup.com/ajcn/article/103/6/1434/4569625?login=

[37] Takahashi S, Hamasuna R, Yasuda M, Arakawa S, Tanaka K, Ishikawa K, et al. A randomized clinical trial to evaluate the preventive effect of cranberry juice (UR65) for patients with recurrent urinary tract infection. Journal of Infection and Chemotherapy. 2013;**19**:112-117. Disponível em: https://pubmed.ncbi.nlm.nih.gov/22961092/

[38] Vasileiou UE, Katsargyris UMA, Theocharis S, Giaginis C. Current clinical status on the preventive effects of cranberry consumption against urinary tract infections. Nutrition Research. 2013;**33**:595-607. https://pubmed.ncbi.nlm.nih.gov/23890348/

[39] Singh I, Gautam LK, Kaur IR. Effect of oral cranberry extract (standardized proanthocyanidin-A) in patients with recurrent UTI by pathogenic E. coli: A randomized placebo-controlled clinical research study. International Urology and Nephrology. 2016;**48**:1379-1386. https://pubmed.ncbi.nlm.nih.gov/27314247/

[40] de Llano G, Liu H, Khoo C, Moreno-Arribas MV, Bartolomé B. Some new findings regarding the antiadhesive activity of cranberry phenolic compounds and their microbial-derived metabolites against Uropathogenic Bacteria. Journal of Agricultural Química Alimentar. 2019;**67**:2166-2174. https://pubs.acs.org/doi/10.1021/acs.jafc.8b05625

[41] Samarasinghe S, Reid R, Al-Bayati M. The anti-virulence effect of cranberry active compound proanthocyanins (PACs) on expression of genes in the third-generation cephalosporin-resistant Escherichia coli CTXM-15 associated with urinary tract infection. 2019. https://www.ncbi.nlm.nih.gov/pmc/articles/PMC6865039/

Section 5

Postharvest Quality Losses of Berry Fruits

Chapter 11

Crop Geometry and Mulch on Strawberry Postharvest Quality

Himadri Shekhar Datta, Pritam Coomar Barua, Utpal Kotoky, Ranjan Das, Hemanta Saikia and Hiranya Devanath

Abstract

Strawberry (*Fragaria* × *ananassa* Duch.) is a natural hybrid species that is cultivated all over the world for its aggregate accessory fruits. Strawberry earns a great respect in the world fruit market owing to its fascinating colour and appealing distinctive flavour, and in India became very popular amongst the farming community due to early production and premium prices. However, the growers do not seem to adopt proper agronomic practices due to various reasons. Amongst the various factors responsible for low production, inappropriate crop geometry and poor selection of mulch material are important. Crop geometry plays a remarkable role for enhancement of strawberry quality through effective utilisation of solar radiation, nutrients and underground resources bringing about better photosynthate formation. The utilisation of mulch in commercial crop production has been practiced to evolve quality strawberry production with reduced disease incidence. The present study was conducted to determine the response of crop geometry and mulch on the postharvest quality of strawberry. Based on yield, quality and economics, the treatment combination (40 cm × 40 cm plant spacing with silver black mulch) was found to be the most viable economic proposition for strawberry in Jorhat condition of Assam.

Keywords: strawberry, spacing, mulch, quality, sensory evaluation, disease incidence

1. Introduction

Strawberry (*Fragaria* × *ananassa* Duch.) belongs to the family Rosaceae. The modern cultivated variety is a hybrid of *Fragaria virginiana* and *Fragaria chiloensis*. The native strawberry, *F. virginiana* was a hardy plant with the ability to withstand cold temperature and drought. It was cultivated by the Early colonists in North America and was imported from North America to Europe in the early 1600s. A wild strawberry, *F. chiloensis*, was found by explorers found in 1700s in Chile. It grew large fruit but was not well suited to a wide range of climates [1].

From the viewpoint of botany, it is aggregate accessory fruit and not berry as fleshy part is derived from central receptacle that holds floral ovary. Imbedded achenes (average 200 on each strawberry) which encompass seeds inside are found in the outermost fruit surface. Owing to its unique organoleptic properties and nutraceutical importance, strawberry is considered as one of the most appealing fruit crops [2].

Strawberry is a non-climacteric and highly perishable fruit. Numerous physiological, morphological, and compositional changes during ripening transform inedible strawberry fruit into a highly cherished fruit. Such changes during the ripening stage include loss of chlorophyll, gain of anthocyanin, increase in sugars, ascorbic acid and pectin, and reduction in acidity, phenolic and cellulose. Also, Disassembly of cell wall mainly due to dissolution of middle lamellae leads to fruit softening at this phase. The fruits which harvested at fully ripened stage are accompanied by high respiration and tissue softening rate, water loss and susceptibility to physical damage which finally leads to fungal infections, particularly Botrytis rot and Rhizopus rot [3]. It holds great importance to adopt appropriate crop management procedures to ensure fruit quality.

The aim of this chapter is to collate and thoroughly narrate the code of crop management practices which need to be followed in the course of pre-harvest operations of strawberry to minimise the losses and certainly fix the quality management concerns.

1. Uses and nutritional benefits: The strawberry fruit has become a functional food providing a number of health benefits apart from basic nutrition as due to its antioxidant, anti-inflammatory, antihyperlipidemic, antihypertensive, or antiproliferative effects affecting modification of aetiology of chronic diseases. Polyphenol and vitamin content contributes for the Antioxidant properties of strawberries with identification of about 40 phenolic compounds, such as, glycosides of quercetin, kaempferol, cyanidin, pelargonidin, ellagic acid, as well as ellagitannins with the most significant contributors being ascorbic acid, ellagitannins and anthocyanins [4].

2. World production: Globally in 2019, a total of 8,885,028 tons of strawberries were produced. Currently, the largest strawberry producer in the world is China followed by United States and Mexico respectively, with a crop area significantly smaller than China but with a yield of more than 50% compared to China. China stands out with 36.2% of the world total, whilst the second and third positions are occupied by the United States and Mexico, with 11.5% and 9.7%, respectively. It is important to note that both Russia and Poland have much larger strawberry area than Mexico, yet they are not close in production levels due to their very low yields. The United States and Mexico have the highest yields worldwide, with 56.3 and 52.4 t/ha, respectively [5].

3. Cultivation and harvesting: Healthy, uniform plants with adequate supply of nutrients contribute towards higher yield of strawberry. Quality planting material and varieties play an important role on the growth and development of any crop. Variation in responses of the varieties to growing practices and the prevailing environment condition during the growing season has been found. Tremendous response of strawberry to major essential elements like N, P and K for its growth, yield and quality has been reported with enhanced marketable yield but an adequate supply is essential for vegetative growth, and desirable yield. Excessive use of these nutrients is not only uneconomical but also induces physiological disorder. Adequate sunshine is prerequisite for strawberry cultivation along with a well-drained soil rich in organic matter. It is advised not to cultivate strawberry in the same land for several years. Alkaline soil or high acidic soil should be avoided with preferred soil pH 5.6 to 6.5. Generation of runner has been found

to be better in light soils with good organic matter. The ideal time for strawberry cultivation is October–November. Land should be well-ploughed and healthy runners are required for planting on raised bed with late afternoon being the best time. It is susceptible to drought as strawberry is relatively shallow-rooted. In absence of frequent irrigation, the plant mortality becomes high. Weed control needs to be taken up in strawberry production. Flowering starts within a month of planting and harvesting can be done within two and half months of its planting. Harvesting done when strawberry ripens and turns bright red [6].

4. Biochemical changes during fruit ripening: Physical, physiological and biochemical changes occurring during fruit ripening seem to modify their internal quality in terms of firmness, colour, starch content, organic acids and flavouring compounds. Strawberry is the model non-climacteric fruit in which with the onset of maturity, the colour of fruits changes due to the accumulation of anthocyanin pigment. Organic acids play a vital role in the growth, and development of fruits and the major organic acid profoundly found in strawberry is ellagic acid. The flavour consists of basically three components viz. aroma, taste and mouth feel. Though fruit taste is contributed by the intricate amalgamation of sugars, organic acids, phenolics and volatile compounds, the distinctive flavour of any specific commodity is attributed to a particular flavouring volatile. The major flavouring compounds identified are esters, alcohols, aldehydes, acids and ketones. Furaneol is a volatile compound found in strawberries and has a role as a flavouring agent, a fragrance and a plant metabolite [7].

Plant spacing is a crop management practice which has a prominent influence on fruit quality but has received a sort of slight attention in strawberry cultivation. Crop management practice like mulch has tremendous influence on crop growth, which finally limits their yield and quality. Strawberry is a surface feeder and therefore mulching plays a very important role in soil moisture conservation, weed control and keeping the fruit clean. Therefore, there is a necessity to adopt appropriate crop management approaches to enhance quality of Strawberry.

2. Crop management practices affecting strawberry fruit quality

2.1 Crop geometry

Amongst the various factors responsible for low production, inappropriate crop geometry is important. Thus, strawberry yield could be increased by using a suitable plant density and establishing an optimum population per unit area of the field is critical to achieving maximum yield. Narrow and wider spacing affects yields because of competition for nutrients, moisture, air, radiation and poor utilisation of the growth factors. With increase in population, yield gets increased proportionally up to a certain level after which the yield declines. Soil fertility, moisture availability, crop growth pattern and cultural practice are some factors affecting spacing [8].

2.2 Mulch

Mulches have a substantial impact on enhancing the sustainable yield and quality of fruit. It improves the physical and chemical qualities of the soil and availability of nutrient pool and biological qualities by increasing beneficial soil microbes. Mulching with different materials significantly increased the physico-chemical qualities of fruits. The practice of applying a layer of dead vegetative waste mulch on soil surface such as straw mulch and polyethylene mulches to conserve soil moisture has been prevalent for a very long time in many areas. Polyethylene mulch colour dictates its energy-radiation behaviour and its effect on the plants' microclimate; it can also affect spectral balance and quantity of canopy light and thus influencing the accumulation of bioactive compounds [9, 10].

3. Effect of crop management practices on percent postharvest disease incidence

Postharvest plant disease can be measured by recording the presence or absence of symptoms known as "incidence", and the degree to which the symptoms are expressed known as "severity". Weather and other environment may cause stress in plants and reduce natural defences by creating favourable conditions for pathogens to infect the plants. Postharvest diseases of fruit can start with fruit set and till harvest. The epidemiology of a disease influenced by weather conditions in the field, and postharvest disease incidence depends on the latent infections initiated in the field during the season. It also depends on fungal propagules contamination during harvest and effectiveness of postharvest treatments including storage and marketing conditions [11].

The time after harvest is characterised by physiological changes in the fruit that favour the occurrence of postharvest diseases, which are often observed only after sale. Postharvest diseases may be typical or quiescent. Pathogens that infect the fruits by wound after harvest cause Typical postharvest diseases. Examples of typical pathogens are *Rhizopus* sp. and *Pestalotiopsis* sp. Pathogens that infect immature fruit prior to harvest cause Quiescent infections. Such infected fruits remain asymptomatic until maturity, and onset of the disease triggered by structural and physiological changes [12]. The interval between harvest and consumption is necessarily short due to the perishability of the fruit. For this reason, many quiescent diseases cannot be observed at the time of product packaging and shipping.

Percent postharvest disease incidence mainly grey mould were recorded in the laboratory by using the formula as mentioned by [13].

Different spacing and mulch treatment combination had significant effect on percent postharvest disease incidence mainly grey mould in strawberry. The strawberry has excellent sensory characteristics but it is highly perishable, with limited postharvest life due to high moisture content, sugars and acids, therefore making it an ideal substrate for proliferation of pathogenic organisms that cause considerable postharvest damage. The pathogens *Botrytis cinerea*, followed by *Rhizopus stolonifer*, *Colletotrichum* spp. *Mucor* spp., are mainly responsible for postharvest diseases of strawberry fruit. *Botrytis cinerea* is the causal organism of grey mould which can cause infection before harvest, in the field stage and stays dormant in storage, till the conditions become favourable for their development [14].

The present study as depicted in **Tables 1** and **2** revealed that the percent postharvest disease incidence was noted significantly least (27.42%) with wider spacing of 40 cm × 40 cm whereas the maximum (31.35%) was recorded with the closest spacing of 20 cm × 30 cm. This might be attributed to the fact that proper plant spacing helps to prevent the development of foliar diseases; because wider rows improve wind

Treatment	Post-harvest disease incidence (%)		
	2019–2020	**2020–2021**	**Pooled**
Spacing (S)			
20 cm × 30 cm	30.79[c]	31.91[c]	31.35[c]
30 cm × 30 cm	29.47[b]	30.63[b]	30.05[b]
30 cm × 40 cm (S_3)	29.07[b]	30.28[b]	29.67[b]
40 cm × 40 cm (S_4)	27.37[a]	28.22[a]	27.79[a]
40 cm × 60 cm (S_5)	26.92[a]	27.93[a]	27.42[a]
SEd (±)	0.41	0.46	0.31
CD (P = 0.05)	0.83	0.94	0.62
Mulches (M)			
Paddy straw (M_1)	31.24[c]	32.32[c]	31.78[c]
Red mulch (M_2)	21.54[b]	22.85[b]	22.20[b]
Silver black mulch (M_3)	20.76[a]	21.99[a]	21.37[a]
No mulch (M_4)	41.36[d]	42.00[d]	41.69[d]
SEd (±)	0.36	0.41	0.28
CD (P = 0.05)	0.74	0.84	0.55
SEd (±)	0.82	0.93	0.62
CD (P = 0.05)	NS	1.88	1.23

Superscript by same letter means they are at par.

Table 1.
Effect of spacing and mulch on post-harvest disease incidence of Strawberry.

Interaction (S × M)	Post-harvest disease incidence (%)						
Treatment combination	**2019–2020**	**2020–2021**	**Pooled**	**Treatment combination**	**2019–2020**	**2020–2021**	**Pooled**
T_1 (S_1M_1)	33.33	34.55	33.94	T_{11} (S_3M_3)	21.25	22.95	22.10
T_2 (S_1M_2)	24.15	25.48	24.81	T_{12} (S_3M_4)	41.34	41.89	41.61
T_3 (S_1M_3)	23.28	24.40	23.84	T_{13} (S_4M_1)	30.46	31.51	30.98
T_4 (S_1M_4)	42.41	43.22	42.82	T_{14} (S_4M_2)	19.32	20.36	19.84
T_5 (S_2M_1)	31.82	32.44	32.13	T_{15} (S_4M_3)	18.80	19.46	19.13
T_6 (S_2M_2)	22.52	24.11	23.32	T_{16} (S_4M_4)	40.89	41.56	41.22
T_7 (S_2M_3)	21.92	23.90	22.91	T_{17} (S_5M_1)	29.30	30.75	30.03
T_8 (S_2M_4)	41.63	42.07	41.85	T_{18} (S_5M_2)	19.29	20.40	19.85
T_9 (S_3M_1)	31.29	32.36	31.82	T_{19} (S_5M_3)	18.55	19.25	**18.90**
T_{10}(S_3M_2)	22.42	23.91	23.16	T_{20} (S_5M_4)	40.56	41.29	40.93

	2019–2020	2020–2021	Pooled
SEd (±)	0.82	0.93	0.62
CD (P = 0.05)	NS	1.88	1.23

The data from the interaction effect of spacing and mulch as depicted in ***Table 2*** *revealed that treatment combination T19 (40 cm × 60 cm with silver black mulch) recorded least percent postharvest disease incidence (18.90%) in the study.*

Table 2.
Effect of spacing and mulch interaction on percent post-harvest disease incidence.

penetration to reduce humidity through the crop canopy, thus making the environment suboptimal for disease infestation. Similar views were expressed by [15] in their study.

Minimum postharvest disease incidence (21.37%) was observed under silver black mulch whereas maximum (41.69%) was recorded under no mulch treatments. It might be due to the fact that the use of plastic mulch may cause less humid surroundings for the plants and, hence, less incidence of Botrytis. Minimum botrytis fruit rot infection observed in plants under silver black polyethylene and maximum in bare soil followed by paddy straw, mainly because straw is favourable medium for spread of *Botrytis cinerea*. The results are in agreement of [16, 17].

The data from the interaction effect of spacing and mulch revealed that treatment combination T19 (40 cm × 60 cm with silver black mulch) recorded least percent postharvest disease incidence (18.90%) in the study. During the entire period of study, the percent disease incidence observed irrespective of the treatments could be attributed by presence of disease causing inoculums due to favourable environmental conditions.

4. Effect of crop management practices on biochemical parameters

Plant density plays an important practice for providing good open position for sunlight, nutrients and moisture availability for quality crop production. To achieve a high-quality yield, high soil moisture is required during the entire growing period. Surface mulches have been used to improve water retention, reduce soil temperature and reduce wind velocity. Several factors such as plant density affect crop biochemical parameters viz. TSS, Sugars, Titratable acidity, Ascorbic acid, fruit pH. TSS can be measured by Zeiss Hand Juice Brix Refractometer and Sugars were estimated by Fehling "A" and "B" solution method as described by [18]. Titratable acidity can be estimated using standard method of [19] and ascorbic acid content estimated by using method of [20]. The pH measurements were made using a digital pH meter and the moisture content of the edible part of fruit estimated by using the method of [21].

Different spacing and mulch treatment combination as presented in **Table 3** had significant effect on quality parameters viz., TSS, total sugar, reducing sugar, non-reducing sugar, titratable acidity, ascorbic acid, fruit juice pH and moisture content in strawberry. Total Soluble Solids (TSS) showed non-significant effect due to spacing and higher TSS (11.39°Brix) noted with wider spacing of 40 cm × 40 cm whereas the lowest TSS (11.16°Brix) was recorded with the closest spacing of 20 cm × 30 cm. This could be due to less interplant competition for light under wider spacing and wider spacing provides larger canopies which provide more photoassimilates, thus increasing TSS. On the other hand, TSS was significantly higher in silver black mulch (12.75°Brix) whilst lowest in no mulch (9.31°Brix). TSS might be more in plastic mulches due to the fact that plastic mulches concentrate CO_2 around the plant canopy and this accumulated CO_2 concentration might have accounted for increased TSS. Similar results were also obtained by [22] who reported higher Total soluble solids (11.96°Brix) under wider spacing of 0.60 m in rockmelon and [23] noticed higher TSS in husk tomato plants from white on black mulching treatment.

Maximum total sugar (6.31%) was observed under widest spacing 40 cm × 60 cm whereas minimum (6.12%) was recorded under closest spacing of 20 cm × 30 cm. Similarly, highest reducing sugar (5.59%) was noted in 40 cm × 40 cm spacing

Treatments	TSS (°Brix)	Total sugar (%)	Reducing sugar (%)	Non reducing sugar (%)	Titratable acidity (%)	Ascorbic acid (mg/100g)	Fruit pH	Moisture content (%)
Spacing (S)								
20 cm × 30 cm (S_1)	11.16	6.12	5.37	0.76	0.74	67.30	3.86	91.57
30 cm × 30 cm (S_2)	11.23	6.18	5.44	0.74	0.74	71.87	3.86	91.70
30 cm × 40 cm (S_3)	11.28	6.24	5.50	0.75	0.73	74.27	3.87	91.74
40 cm × 40 cm (S_4)	11.39	6.30	5.55	0.76	0.72	77.61	3.91	91.70
40 cm × 60 cm (S_5)	11.36	6.31	5.59	0.72	0.71	78.65	3.88	91.51
SEd (±)	0.10	0.01	0.01	0.01	0.02	1.47	0.01	0.22
CD (P = 0.05)	—	0.02	0.02	—	—	2.92	—	—
Mulches (M)								
Paddy straw (M_1)	11.06	6.15	5.42	0.74	0.80	65.73	3.84	92.05
Red mulch (M_2)	12.02	6.31	5.52	0.78	0.65	82.67	3.92	91.04
Silver black mulch (M_3)	12.75	6.38	5.61	0.77	0.65	85.81	3.93	91.09
No mulch (M_4)	9.31	6.08	5.39	0.69	0.83	61.53	3.81	92.40
SEd (±)	0.09	0.01	0.01	0.01	0.02	1.31	0.01	0.19
CD (P = 0.05)	0.18	0.02	0.02	0.02	0.03	2.61	0.02	0.38

Table 3.
Effect of spacing and mulch on quality of strawberry.

whereas minimum reducing sugar (5.37%) was observed in 20 cm × 30 cm. Sugar content was recorded higher under wider spacing possibly due to higher photosynthesis and availability of metabolites because of higher interception of Photosynthetically active radiation (PAR) by individual plant and better translocation and accumulation of nutrients. Maximum total sugar (6.38%) and reducing sugar (5.61%) under silver black mulch might be attributed to high soil temperature and higher nutrient availability which provided favourable microclimate for increased sugar content. The current findings are in consonance with results of [24] who got total sugars (8.50%), reducing sugar (4.78%) in strawberry under wider spacing and [25] reported total sugar (5.71%), reducing sugar (3.41%), non-reducing sugar (2.30%) in aonla.

Titratable acidity was found to have non-significant effect due to spacing. Higher titratable acidity (0.74%) was obtained under closer spacing of 20 cm × 30 cm and 30 cm × 30 cm whereas minimum titratable acidity (0.71%) was recorded under wider spacing of 40 cm × 60 cm which might be due to shade effect where sugar conversion from organic acid is hampered due to lack of sufficient light in closer spacing. Least titratable acidity (0.65%) was found in silver black mulch and red mulch which might be due to rapid conversion of organic acid to sugars and more reflection of photosynthetically active radiation(PAR) from plastic mulch into fruiting zone elevated conversion into sugars and reduction of acidity [24] obtained lowest titrable acidity (0.812%) under wider spacing of 50 cm × 40 cm and [26] reported least acidity (1.40%) in plants under black polyethylene film and higher acid content (1.52%) under unmulched plants during his study on plum.

Perusal of data revealed that the maximum ascorbic acid (78.65 mg 100 g^{-1}) was recorded in 40 cm × 60 cm which was at par with 40 cm × 40 cm (77.61 mg 100 g^{-1}) whilst the minimum ascorbic acid was recorded in 20 cm × 30 cm (67.30 mg 100 g^{-1}). Higher ascorbic acid in wider spacing might be due to more light exposure and greater accumulation of photosynthates. Ascorbic acid content was significantly influenced by mulch treatments, with maximum ascorbic acid (85.81 mg 100 g^{-1}) recorded in silver black mulch whereas no mulch reported minimum ascorbic acid (61.53 mg 100 g^{-1}). Improvements in ascorbic acid in polythene treatments may be due to promotion effect of plant growth and metabolic processes which led to increase in chemical composition. Reflection of light from polythene mulches might have played a role in modifying microclimatic conditions. The results are in conformity with findings of [27] who obtained higher ascorbic acid (178.78 mg 100 g^{-1}) in the wider spacing of 6 m × 6 m and [28] found higher ascorbic acid (26.64 mg 100 g^{-1}) under silver black mulch in tomato compared to paddy straw (25.30 mg 100 g^{-1}) and no mulch treatments (24.28 mg 100 g^{-1}).

In the present study, no significant effect of spacing on fruit pH was found. The increase in pH could be related to a possible decrease in the respiratory metabolic activity or increase in ascorbic acid content as expressed by [29]. Higher fruit pH (3.93) under silver black mulch could be due to changes in quality and quantity of light energy re-radiated into plant canopy from mulch that might have influenced fruit pH in agreement with [30] who got higher juice pH (3.57) in pomegranate plants mulched with black polythene.

Effect of spacing on fruit moisture content was non-significant whereas in red mulch treatment, minimum fruit moisture content (91.04%) was recorded that could be due to better conversion of starch into sugars. Hence lower relative water content in fruits with increase in ripeness level in agreement with views of [31].

The strawberry fruits lying on the ground come in contact with soil making the fruits dirty and susceptible to soil-borne pathogen infections, leading to reduced fruit quality. This is an additional advantage of using mulches in strawberry production by reducing the number of diseased and dirty berries [32].

5. Effect of crop management practices on sensory evaluation

The fruits of different treatments were subjected to sensory evaluation for evaluating firmness, skin colour, pulp colour, flavour and sweetness are presented in **Table 4**.

The highest numerical rating with respect to firmness (Hedonic scale rating 4.14), skin colour (4.42), pulp colour (4.28), flavour (3.57), sweetness (3.71) was observed under treatment combination T15 (40 cm × 40 cm spacing with silver black mulch) having good overall acceptability. The numerical rating was lowest with respect to firmness (3.28), skin colour (3.28), pulp colour (2.28), flavour (2.00), sweetness (1.85) were recorded in T17 (40 cm × 60 cm spacing and paddy straw) giving acceptability score of 2.53 and T12 (30 cm × 30 cm with no mulch) giving score of 2.93 and both had the overall acceptability below 3.

Treatment combination	Sensory evaluation rating					
	Firmness	Skin colour	Pulp colour	Flavour	Sweetness	Average
T_1 (S_1M_1)	3.71	3.85	2.57	2.85	2.71	3.13
T_2 (S_1M_2)	3.42	4.42	3.71	3.85	3.57	3.79
T_3 (S_1M_3)	3.71	4.42	3.57	3.57	3.57	3.77
T_4 (S_1M_4)	3.57	4.57	3.71	3.42	3.28	3.71
T_5 (S_2M_1)	3.85	4.28	3.28	3.42	2.85	3.53
T_6 (S_2M_2)	3.85	4.71	3.71	3.42	3.14	3.76
T_7 (S_2M_3)	3.42	4.14	3.57	3.57	2.42	3.42
T_8 (S_2M_4)	3.57	4.00	3.71	3.42	3.14	3.56
T_9 (S_3M_1)	3.57	3.57	3.57	3.14	3.42	3.45
T_{10} (S_3M_2)	3.85	4.71	4.00	3.42	3.42	3.88
T_{11} (S_3M_3)	3.85	3.71	3.14	3.00	2.71	3.28
T_{12} (S_3M_4)	3.42	3.28	2.57	3.00	2.42	2.93
T_{13} (S_4M_1)	3.42	3.57	3.00	3.71	3.42	3.42
T_{14} (S_4M_2)	3.85	4.00	3.14	3.14	3.00	3.42
T_{15} (S_4M_3)	4.14	4.42	4.28	3.57	3.71	**4.02**
T_{16} (S_4M_4)	3.85	4.00	3.57	3.28	3.14	3.56
T_{17} (S_5M_1)	3.28	3.28	2.28	2.00	1.85	2.53
T_{18} (S_5M_2)	3.71	3.85	3.85	3.57	3.14	3.62
T_{19} (S_5M_3)	3.00	4.28	3.71	3.14	2.42	3.31
T_{20} (S_5M_4)	3.85	4.42	4.00	3.14	3.00	3.68

The fruits of different treatments were subjected to sensory evaluation and the effect of spacing and mulch interaction on firmness, skin colour, pulp colour, flavour and sweetness presented in ***Table 4.***

Table 4.
Effect of spacing and mulch interaction on sensory evaluation of Strawberry.

The better organoleptic qualities under wider spacing could be due to fact that sufficient sunlight under wider spacing might have led to better conversion of starch to sugar as observed by [33]. The better organoleptic traits under plastic mulch treatment could be due to more reflection of photosynthetically active radiation (PAR) into fruiting zone which elevated sugar conversion in agreement with views of [34]. In the study, the panel gave at par ratings to the samples for acceptance of the external appearance, indicating that the consumers would accept better sample, despite differences in physical characteristics. The average hedonic value, however, did not always faithfully represent the opinions of the judges as a group as opined by [35].

6. Conclusion

Strawberry is a highly perishable fruit and there are several crop management practices which have effect on fruit quality. Numerous crop management strategies such as spacing and mulching need to be scaled up to enhance the fruit quality.

The study indicated that fruits grown under 40 cm × 40 cm spacing with silver black mulch had least percent postharvest disease incidence along with better organoleptic traits and quality of fruits in strawberry.

Acknowledgements

The author would like to acknowledge Assam Agricultural University, Jorhat for providing necessary guidance and help during the course of study. Sincere gratitude to the farmer for providing all support for conducting the experiment in field. Moreover, we are thankful to the reviewers for their critical reading and suggestions for the improvement of our manuscript.

Conflict of interest

Authors have declared that no conflict of interests exists.

Author details

Himadri Shekhar Datta*, Pritam Coomar Barua, Utpal Kotoky, Ranjan Das, Hemanta Saikia and Hiranya Devanath
Assam Agricultural University, Jorhat, India

*Address all correspondence to: himadri.shekhar.datta@aau.ac

References

[1] Husaini AM, Zaki FA. Strawberries: A general account. In: Husaini AM, Neri D, editors. Strawberry: Growth, Development and Diseases. Wallingford, Oxfordshire, England; CAB International; 2016. pp. 1-9

[2] Giampieri F, Tulipani S, Alvarez Suarez JM, Quiles JL, Mezzetti B, Battino M. The strawberry: Composition, nutritional quality, and impact on human health. Nutrition. 2012;**28**:9-19

[3] Azam M, Ejaz S, Rehman RN, Khan M, Qadri R. Postharvest quality management of strawberries. In: Asao T, Asaduzzaman M, editors. Strawberry—Pre- and Post-Harvest Management Techniques for Higher Fruit Quality. London, UK: Intech Open; 2019. pp. 1-21

[4] Basu A, Nguyen A, Betts NM, Lyons TJ. Strawberry As a functional food: An evidence-based review. Critical Reviews in Food Science and Nutrition. 2014;**54**:790-806

[5] López Y, Olmo Axayacatl MC. Strawberry analysis 2021: Production, Prices & Covid impact. Treembo. 2021;**2021**:1-67

[6] Mondal, T. , Ali, M.S., Bagum,T, Das, D.and Islam, M.M. Use of improved practices in strawberry cultivation by the farmers in selected area of Bangladesh. International Journal of Business, Social and Scientific Research. 2016;**4**(3):226-230

[7] Gundewadi G, Reddy VR, Bhimappa BB. Physiological and biochemical basis of fruit development and ripening—A review. Journal of Hill Agriculture. 2018;**9**(1):7-21

[8] Bhattarai P, Tripathi KM, Gautam DM, Shrestha AK. Effect of crop geometry on growth, yield and quality of sweet potato (Ipomoea batatas L.) genotypes. Turkish Journal of Agriculture - Food Science and Technology. 2022;**10**(8):1532-1541

[9] Das K, Dutta P. Effects of mulching on soil properties and post-harvest quality of Mango Cv. Himsagar Grown in New Alluvial Zone of West Bengal, International Journal of Agriculture, Environment and Biotechnology. 2018;**11**(2):259-264

[10] Mlaba MD, Fotouo-M H, Tesfay SZ, Odindo A, Magwaza LS, Mditshwa A, et al. Effect of polyethylene mulch colour and maturity stage on strawberry phytochemicals and postharvest quality. Acta Horticulturae. 2022;**2022**(1349):251-260

[11] Michailides TJ, Morgan DP, Luo Y. Epidemiological assessments and postharvest disease incidence. In: Prusky D, Gullino ML, editors. Postharvest Pathology, Plant Pathology in the 21st Century. Vol. 2nd. Springer Dordrecht, Heidelberg, London, New York: Springer Science; 2009. pp. 69-88

[12] Barkai-Golan R. Postharvest Diseases of Fruits and Vegetables: Development and Control. Amsterdam: The Netherlands. Elsevier Science; 2001

[13] Kuyu CG, Tola YB. Assessment of banana fruit handling practices and associated fungal pathogens in Jimma town market, Southwest Ethiopia. Food Science & Nutrition. 2018;**2018**:1-8

[14] Datta Himadri S, Barua PC, Kotoky U, Das R, Saikia H, Deva Nath HK. Effect of spacing and mulch on post harvest disease incidence and

sensory evaluation in strawberry. Asian Journal of Agril Extension, Economics and Sociology. 2022;**40**(9): 47-52

[15] Brar JS, Dhaliwal HS, Bal JS, Singh SP. Influence of inter-plant spacing on microclimatic and horticultural parameters of guava (*Psidium guajava* L.) orchard. Research. Crops. 2013;**14**(1):205-212

[16] Kumar PS, Choudhary VK, Bhagawati R. Influence of mulching and irrigation level on water-use efficiency, plant growth and quality of strawberry (Fragaria x ananassa). Indonesia Journal of Agricultural Science. 2012;**82**(2):127-133

[17] Sharma RR, Sharma VP, Pandey SN. Mulching influences plant growth and albinism disorder in strawberry under subtropical climate. Acta Horticulturae. 2004;**662**:187-191

[18] Ranganna S. Manual of Analysis of Fruit and Vegetable Products. New Delhi: Tata: McGraw-Hill; 1977

[19] AOAC. In: Horwitz W, editor. Official Methods of Analysis of the Association of Analytical Chemists. Washington D.C.: AOAC; 1980. p. 1018

[20] Freed M. Methods of Vitamin Assay. New York: Interscience Publishers; 1966

[21] AOAC. Official methods of analysis of the Association of Analytical Chemists. In: Official Method of Analysis. 17th ed. Washington D.C; 2000. p. 1018

[22] Wee WC, Lai K, Kong CL, Yap WS. Impact of within-row plant spacing and fixed fruit setting on yield and quality of Rockmelon fruit cultivated by drip irrigation in a greenhouse. Horticultural Science and Technology. 2018;**36**(2):172-182

[23] Helaly AA, Goda Y, Abd El-Rehim AS, Mohamed AA, El-Zeiny OAH. Effect of polyethylene mulching type on the growth, yield and fruits quality of Physalis pubescens. Advanced Plants Agricultural Research. 2017;**6**(5):154-160

[24] Hazarika TK, Lalrinfeli L, Lalchhanmawia J, Nautiyal BP. Effect of planting density on growth, yield and quality of strawberry (Fragaria × ananassa) cv. Camarosa Indian Journal of Agricultural Science. 2019;**89**(3):489-493

[25] Iqbal M, Bakshi P, Kumar R, Wali VK, Bhushan B. Influence of mulching on fruit quality of aonla (Emblica officinalis Gaertn.) cv. Ecosystem Environment and Conservation. 2015;**21**(3):263-268

[26] Mehraj S, Qurtulane FA, Peer IA, Bisati MM, Hassan S. Effect of organic and inorganic mulches on TSS, acidity, TSS/acidity, yield efficiency, soil hydrothermal conditions and leaf nutrient content under temperate conditions. Ecosystem Environmental & Conservation. 2015;**21**(3):1331-1337

[27] Bhagyashree M, Naik N, Koujalagi CB, Karadiguddi M, Manukumar HR. Effect of plant geometry and nutrients on nutrient status and fruit quality of guava (Psidium guajava L) cv. Sardar International Journal of Chemical Studies. 2018;**6**(4):3223-3227

[28] Gad Moula MMH, Sadek II, Moursy FS. Effect of plastic color and organic mulching on the yield of tomato and lettuce. International Journal of Science Research Method. 2018;**9**(2):173-191

[29] Da Silva EP, Cardoso AFL, Fante C, Rosell CM, Vilas Boas EVB. Effect of

postharvest temperature on the shelf life of gabiroba fruit (Campomanesia pubescens). Food Science Technology. 2013;**33**(4):632-637

[30] Yogaraj S. Studies on the Influence of Different Types of Mulch on Growth, Yield and Quality of Pomegranate (Punica granatum L) cv. Bhagwa. Bagalkot: Univ of Horti. Sci; 2016

[31] Frenkel C, Hartman TG. Decrease in fruit moisture content heralds and might launch the onset of ripening processes. Journal of Food Science. 2012;77:S365-S376

[32] Reham MA, Attalah SY, Aboalmajd BIFM, SA. Effect of plastic mulching on strawberry fruit yield and quality. Assiut. The Journal of Agricultural Science. 2019;**50**(4):126-141

[33] Anil GY. Studies on Plant Density and Nutrient Requirement in Banana cv.Grand Naine.PhD Thesis Submitted to V.N.M. Parbhani: Krishividyapeeth; 2019

[34] Dhanush KV, Patil DR, Kantesh G, Patil SN, Alur A. Studies on the influence of different types of studies on the influence of different types of mulches on growth, yield and quality of Grapes (Vitis vinifera L.) cv. Kishmish Rozavis White. Bioscan. 2017;**12**(2):953-957

[35] Amorim AV, Garruti DS, Lacerda CF, Moura CFH, Gomes-Filho E. Postharvest and sensory quality of pineapples grown under micronutrients doses and two types of mulching. African Journal of Agricultural Research. 2013;**8**(19):2240-2224